Compendium of Plant Genomes

Series Editor

Chittaranjan Kole, President, International Climate Resilient Crop Genomics Consortium (ICRCGC), President, International Phytomedomics and Nutriomics Consortium (IPNC) and President, Genome India International (GII), Kolkata, India

Whole-genome sequencing is at the cutting edge of life sciences in the new millennium. Since the first genome sequencing of the model plant *Arabidopsis thaliana* in 2000, whole genomes of about 100 plant species have been sequenced and genome sequences of several other plants are in the pipeline. Research publications on these genome initiatives are scattered on dedicated web sites and in journals with all too brief descriptions. The individual volumes elucidate the background history of the national and international genome initiatives; public and private partners involved; strategies and genomic resources and tools utilized; enumeration on the sequences and their assembly; repetitive sequences; gene annotation and genome duplication. In addition, synteny with other sequences, comparison of gene families and most importantly potential of the genome sequence information for gene pool characterization and genetic improvement of crop plants are described.

K. K. Sabu · Chittaranjan Kole
Editors

The Cardamom Genome

Springer

Editors
K. K. Sabu
Jawaharlal Nehru Tropical Botanic
Garden and Research Institute
Thiruvananthapuram, Kerala, India

Chittaranjan Kole
Prof. Chittaranjan Kole Foundation
for Science and Society
New Delhi, Delhi, India

ISSN 2199-4781 ISSN 2199-479X (electronic)
Compendium of Plant Genomes
ISBN 978-3-031-71828-1 ISBN 978-3-031-71829-8 (eBook)
https://doi.org/10.1007/978-3-031-71829-8

This Springer imprint is published by the registered company Springer Nature Switzerland AG
The registered company address is: Gewerbestrasse 11, 6330 Cham, Switzerland

This book series is dedicated to my wife Phullara
and our children Sourav and Devleena

Chittaranjan Kole

Preface to the Series

Genome sequencing has emerged as the leading discipline in the plant sciences coinciding with the start of the new century. For much of the twentieth century, plant geneticists were only successful in delineating putative chromosomal location, function, and changes in genes indirectly through the use of a number of "markers" physically linked to them. These included visible or morphological, cytological, protein, and molecular or DNA markers. Among them, the first DNA marker, the RFLPs, introduced a revolutionary change in plant genetics and breeding in the mid-1980s, mainly because of their infinite number and thus potential to cover maximum chromosomal regions, phenotypic neutrality, absence of epistasis, and codominant nature. An array of other hybridization-based markers, PCR-based markers, and markers based on both facilitated construction of genetic linkage maps, mapping of genes controlling simply inherited traits, and even gene clusters (QTLs) controlling polygenic traits in a large number of model and crop plants. During this period, a number of new mapping populations beyond F_2 were utilized and a number of computer programs were developed for map construction, mapping of genes, and for mapping of polygenic clusters or QTLs. Molecular markers were also used in the studies of evolution and phylogenetic relationship, genetic diversity, DNA fingerprinting, and map-based cloning. Markers tightly linked to the genes were used in crop improvement employing the so-called marker-assisted selection. These strategies of molecular genetic mapping and molecular breeding made a spectacular impact during the last one-and-a-half decades of the twentieth century. But still they remained "indirect" approaches for elucidation and utilization of plant genomes since much of the chromosomes remained unknown and the complete chemical depiction of them was yet to be unraveled.

Physical mapping of genomes was the obvious consequence that facilitated the development of the "genomic resources" including BAC and YAC libraries to develop physical maps in some plant genomes. Subsequently, integrated genetic–physical maps were also developed in many plants. This led to the concept of structural genomics. Later on, emphasis was laid on EST and transcriptome analysis to decipher the function of the active gene sequences leading to another concept defined as functional genomics. The advent of techniques of bacteriophage gene and DNA sequencing in the 1970s was extended to facilitate sequencing of these genomic resources in the last decade of the twentieth century.

As expected, sequencing of chromosomal regions would have led to too much data to store, characterize, and utilize with the-then available computer software could handle. But the development of information technology made the life of biologists easier by leading to a swift and sweet marriage of biology and informatics, and a new subject was born—bioinformatics.

Thus, the evolution of the concepts, strategies, and tools of sequencing and bioinformatics reinforced the subject of genomics—structural and functional. Today, genome sequencing has traveled much beyond biology and involves biophysics, biochemistry, and bioinformatics!

Thanks to the efforts of both public and private agencies, genome sequencing strategies are evolving very fast, leading to cheaper, quicker, and automated techniques right from clone-by-clone and whole-genome shotgun approaches to a succession of second-generation sequencing methods. The development of software for different generations facilitated this genome sequencing. At the same time, newer concepts and strategies were emerging to handle sequencing of the complex genomes, particularly the polyploids.

It became a reality to chemically—and so directly—define plant genomes, popularly called whole-genome sequencing or simply genome sequencing.

The history of plant genome sequencing will always cite the sequencing of the genome of the model plant Arabidopsis thaliana in 2000 that was followed by sequencing the genome of the crop and model plant rice in 2002. Since then, the number of sequenced genomes of higher plants has been increasing exponentially, mainly due to the development of cheaper and quicker genomic techniques and, most importantly, the development of collaborative platforms such as national and international consortia involving partners from public and/or private agencies.

As I write this preface for the first volume of the new series "Compendium of Plant Genomes," a net search tells me that complete or nearly complete whole-genome sequencing of 45 crop plants, eight crop and model plants, eight model plants, 15 crop progenitors and relatives, and three basal plants is accomplished, the majority of which are in the public domain. This means that we nowadays know many of our model and crop plants chemically, i.e., directly, and we may depict them and utilize them precisely better than ever. Genome sequencing has covered all groups of crop plants. Hence, information on the precise depiction of plant genomes and the scope of their utilization are growing rapidly every day. However, the information is scattered in research articles and review papers in journals and dedicated Web pages of the consortia and databases. There is no compilation of plant genomes and the opportunity of using the information in sequence-assisted breeding or further genomic studies. This is the underlying rationale for starting this book series, with each volume dedicated to a particular plant.

Plant genome science has emerged as an important subject in academia, and the present compendium of plant genomes will be highly useful to both students and teaching faculties. Most importantly, research scientists involved in genomics research will have access to systematic deliberations on the plant genomes of their interest. Elucidation of plant genomes is of interest not only for the geneticists and breeders, but also for practitioners of an array of plant science disciplines, such as taxonomy, evolution, cytology, physiology, pathology, entomology, nematology, crop production, biochemistry, and obviously bioinformatics. It must be mentioned that information regarding each plant genome is ever-growing. The contents of the volumes of this compendium are, therefore, focusing on the basic aspects of the genomes and their utility. They include information on the academic and/or economic importance of the plants, description of their genomes from a molecular genetic and cytogenetic point of view, and the genomic resources developed. Detailed deliberations focus on the background history of the national and international genome initiatives, public and private partners involved, strategies and genomic resources and tools utilized, enumeration on the sequences and their assembly, repetitive sequences, gene annotation, and genome duplication. In addition, synteny with other sequences, comparison of gene families, and, most importantly, the potential of the genome sequence information for gene pool characterization through genotyping by sequencing (GBS) and genetic improvement of crop plants have been described. As expected, there is a lot of variation of these topics in the volumes based on the information available on the crop, model, or reference plants.

I must confess that as the series editor, it has been a daunting task for me to work on such a huge and broad knowledge base that spans so many diverse plant species. However, pioneering scientists with lifetime experience and expertise on the particular crops did excellent jobs editing the respective volumes. I myself have been a small science worker on plant genomes since the mid-1980s and that provided me the opportunity to personally know several stalwarts of plant genomics from all over the globe. Most, if not all, of the volume editors are my longtime friends and colleagues. It has been highly comfortable and enriching for me to work with them on this book series. To be honest, while working on this series I have been and will remain a student first, a science worker second, and a series editor last. And, I must express my gratitude to the volume editors and the chapter authors for providing me the opportunity to work with them on this compendium.

I also wish to mention here my thanks and gratitude to Springer staff, particularly Dr. Christina Eckey and Dr. Jutta Lindenborn, for the earlier set of volumes and presently Ing. Zuzana Bernhart for all their timely help and support.

I always had to set aside additional hours to edit books besides my professional and personal commitments—hours I could and should have given to my wife, Phullara, and our kids, Sourav and Devleena. I must mention that they not only allowed me the freedom to take away those hours from them but also offered their support in the editing job itself. I am really not sure whether my dedication of this compendium to them will suffice to do justice to their sacrifices for the interest of science and the science community.

New Delhi, India Chittaranjan Kole

genomes of their interest. Elucidation of plant genomes is of interest not only for the geneticists and breeders, but also for practitioners of an array of plant science disciplines, such as taxonomy, evolution, cytology, physiology, pathology, entomology, nematology, crop production, biochemistry, and obviously bioinformatics. It must be mentioned that information regarding each plant genome is ever-expanding. The contents of the volumes of this compendium are, therefore, focusing on the basic aspects of the genomes and their utility. [illegible]

[illegible]

[illegible] ... would like to take this opportunity to thank ... [illegible] (my wife, Phullara, and our kids Sourav and Devleena). I must mention that they not only allowed me the freedom to take away those hours from them but also offered their support in the editing job itself. I am really not sure whether my dedication of this compendium to them will suffice to do justice to their feelings. For the interest of science and the science community.

New Delhi, India Chittaranjan Kole

Preface

The production of the cardamom Genome book is essential and highlights the groundbreaking research ongoing for this critical spice crop. This volume includes chapters detailing cardamom as a leading spice crop for food flavor and health, genetic resources, genetics and breeding, in vitro culture, genomics of biotic stress tolerance, volatiles, transcriptomic resources and their implications, molecular characterization and phylogenetic relationships, functional regulation of miRNAs and whole-genome sequence and SSR markers. We are indebted to our colleagues from the global cardamom community who contributed to this book and for their continuous commitment to provide resources and knowledge to scientists and breeders around the world.

Thiruvananthapuram, India — K. K. Sabu
Kolkata, India — Chittaranjan Kole

Preface

About this Book

This book provides the first comprehensive coverage of the cardamom genome sequence. The book provides a broad and extensive review of the resources, tools, and methodologies available for exploiting the cardamom genome sequence for crop improvement and studying fundamental questions related to the structure, function, and genetic resources of the cardamom. Cardamom, native to India and Sri Lanka, is believed to have its root origin in the tropical evergreen woods of the Western Ghats in South India and is extensively in cultivation globally. This book is useful to students, teachers, and scientists in academia and industry who are interested in gaining an understanding of the cardamom genome and its application.

Contents

Contents

Contributors

H. J. Akshitha ICAR-Indian Institute of Spices Research, Regional Station, Appangala, Madikeri, Kodagu (D), Karnataka, India

H. N. Anjana Biotechnology and Bioinformatics Division, Jawaharlal Nehru Tropical Botanic Garden and Research Institute (JNTBGRI), Palode, Thiruvananthapuram, India

S. J. Ankegowda ICAR-Indian Institute of Spices Research, Regional Station, Appangala, Madikeri, Kodagu (D), Karnataka, India

Sharon Aravind ICAR-Indian Institute of Spices Research, Kozhikode, Kerala, India

Honnappa Asangi ICAR-Indian Institute of Spices Research, Regional Station, Appangala, Madikeri, Kodagu (D), Karnataka, India

M. Balaji Rajkumar ICAR-Indian Institute of Spices Research, Regional Station, Appangala, Madikeri, Kodagu (D), Karnataka, India

M. Balakumaran ICAR—Indian Institute of Spices Research, Kozhikode, India

J. Berliner ICAR—Indian Agricultural Research Institute, Nilgiris, India

Ana Cristina Tamayo Domínguez EARTH University, Guácimo, Costa Rica

Sangeetha Elangovan Department of Plant Sciences, School of Biological Sciences, Madurai Kamaraj University, Madurai, Tamil Nadu, India; Department of Biotechnology, Thiagarajar College, Madurai, Tamil Nadu, India

Ambika Baldev Gaikwad Division of Genomic Resources, ICAR-National Bureau of Plant Genetic Resources, New Delhi, India

R. Gobu ICAR—Sugarcane Breeding Research Institute, Coimbatore, India

Julius Jeyaprakasam Department of Plant Sciences, School of Biological Sciences, Madurai Kamaraj University, Madurai, Tamil Nadu, India

Ratna Kumari Division of Genomic Resources, ICAR-National Bureau of Plant Genetic Resources, New Delhi, India

K. Mary Mathew Indian Cardamom Research Institute—ICRI, Spices Board, GoI, Kochi (Retired), Thrissur, Kerala, India

P. Mohammed Faisal ICAR-Indian Institute of Spices Research, Regional Station, Appangala, Madikeri, Kodagu (D), Karnataka, India

F. Nadiya Inter-University Centre for Genomics and Gene Technology, University of Kerala, Karyavattom, Kerala, India

D. Prasath ICAR-Indian Institute of Spices Research, Kozhikode, Kerala, India

P. M. Priyadarshan (formerly with the Rubber Research Institute of India), Thiruvananthapuram, India

R. Rajalakshmi Inter-University Centre for Genomics and Gene Technology, University of Kerala, Karyavattom, Kerala, India;
Department of Botany, University of Kerala, Karyavattom, Kerala, India

Y. S. Rao Indian Cardamom Research Institute ICRI, Spices Board, GoI, Kochi (Retired), Kochi, India

K. K. Sabu KSCSTE—Jawaharlal Nehru Tropical Botanic Garden and Research Institute, Palode, Thiruvananthapuram, Kerala, India;
Biotechnology and Bioinformatics Division, Jawaharlal Nehru Tropical Botanic Garden and Research Institute (JNTBGRI), Palode, Thiruvananthapuram, India

C. Sellaperumal ICAR—Indian Institute of Spices Research, Kozhikode, India

M. S. Shivakumar ICAR-Indian Institute of Spices Research, Regional Station, Appangala, Madikeri, Kodagu (D), Karnataka, India

R. Sivaranjani ICAR-Indian Institute of Spices Research, Kozhikode, Kerala, India

S. D. Soumya Biotechnology and Bioinformatics Division, Jawaharlal Nehru Tropical Botanic Garden and Research Institute (JNTBGRI), Palode, Thiruvananthapuram, India

Jebasingh Tennyson Department of Plant Sciences, School of Biological Sciences, Madurai Kamaraj University, Madurai, Tamil Nadu, India

Sheel Yadav Division of Genomic Resources, ICAR-National Bureau of Plant Genetic Resources, New Delhi, India

Abbreviations

2,4-D	2,4-Dichlorophenoxyacetic Acid
Acc.	Accession
AD	Anno Domini
AFLP	Amplified Fragment Length Polymorphism
AGO1	ARGONAUTE1
AICRPS	All India Coordinated Research Project on Spices
APG	Angiosperm Phylogeny Group
ARF	Auxin Response Factor
BAP	6-Benzylaminopurine
BC1	Backcross 1
bHLH	Basic Helix-Loop-Helix
BLAST	Basic Local Alignment Search Tool
BUSCO	Benchmarking Universal Single-Copy Orthologs
CAFÉ	Computational Analysis of gene Family Evolution
CAGR	Compound Annual Growth Rate
CdMV	Cardamom Mosaic Virus
CeNAT	Centro Nacional de Alta Tecnología
CI	Cytoplasmic Inclusion
CP	Coat Protein
cpSSR	Chloroplast SSR
CTAB	Cetyl Trimethyl Ammonium Bromide
CV	Coefficient of Variation
DBT	Department of Biotechnology
DEGs	Differentially Expressed Genes
EST-SSR	Expressed Sequence Tag-Simple Sequence Repeats
F1	First Filial Generation
F2	Second Filial Generation
FGB	Not specified in the document; assumed to be a genotype or line identifier
Gb	Gigabases
GBIF	Global Biodiversity Information Facility
GCA	General Combining Ability
GC-MS	Gas Chromatography-Mass Spectrometry
GO	Gene Ontology
GSPC	Global Strategy for Plant Conservation

HC-Pro	Helper Component Proteinase
HSP	Heat Shock Protein
IAA	Indole-3-Acetic Acid
ICAR	Indian Council of Agricultural Research
ICRI	Indian Cardamom Research Institute
IIHR	Indian Institute of Horticultural Research
IISR	Indian Institute of Spices Research
Indels	Insertions and Deletions
IPGRI	International Plant Genetic Resources Institute
ISSR	Inter-Simple Sequence Repeat
ITS	Internal Transcribed Spacer
JNTBGRI	Jawaharlal Nehru Tropical Botanic Garden and Research Institute
KEGG	Kyoto Encyclopedia of Genes and Genomes
KOG	Eukaryotic Orthologous Groups
LAT	Linalool Acetyl Transferase
LRR-RLK	Leucine-Rich Repeat—Receptor-Like Kinases
LTR	Long Terminal Repeat
MF	Molecular Functions
miRNA	MicroRNA
mRNA	Messenger RNA
MS	Murashige and Skoog Medium
MSL	Mean Sea Level
MT	Metric Ton
mtSSR	Mitochondrial SSR
MVSP	MultiVariate Statistical Package
MYB	Myeloblastosis a family of transcription factors
NAA	Naphthalene Acetic Acid
NBS-LRR	Nucleotide Binding Site—Leucine-Rich Repeat
NCL	National Chemical Laboratory
NGS	Next-Generation Sequencing
NIa	Nuclear Inclusion a
NIb	Nuclear Inclusion b
NKE	Not specified in the document; assumed to be a genotype or line identifier
NR	Non-redundant
nuSSR	Nuclear SSR
ORF	Open Reading Frame
PCR	Polymerase Chain Reaction
PCR-RFLP	Polymerase Chain Reaction-Restriction Fragment Length Polymorphism
PGR	Plant Growth Regulator
PIPO	Pretty Interesting Potyviridae ORF
PNG	Papua New Guinea
Pri-miRNAs	Primary MicroRNAs
RAPD	Random Amplified Polymorphic DNA
RFLP	Restriction Fragment Length Polymorphism

RGA	Resistance Gene Analog
RISC	RNA-Induced Silencing Complex
RLK	Receptor-Like Kinases
RNA	Ribonucleic Acid
rRNA	Ribosomal RNA
SCAR	Sequence-Characterized Amplified Region
SCFE	Supercritical Fluid Extraction
SCMVTDb	Small Cardamom Mosaic Virus Transcriptome Database
SE	Somatic Embryogenesis
SmCarTDB	Small Cardamom Transcriptome Database
SNPs	Single Nucleotide Polymorphisms
SSR	Simple Sequence Repeats
TAIR	The Arabidopsis Information Resource
TC	Tissue Culture
TFs	Transcription Factors
TPL	The Plant List
tRNA	Transfer RNA
UPGMA	Unweighted Pair Group Method with Arithmetic Mean
UPOV	International Union for the Protection of New Varieties of Plants
VPg	Viral Protein Genome-linked
WFO	World Flora Online
WGS	Whole-Genome Sequencing

Cardamom: A Leading Spice Crop for Food, Flavor, and Health

1

H. J. Akshitha, Sharon Aravind, M. S. Shivakumar, D. Prasath, S. J. Ankegowda, P. Mohammed Faisal, Honnappa Asangi, and M. Balaji Rajkumar

Abstract

In India, cardamom is one among the important spice crops cultivated. It is one of the most consumed and traded commodities in the world. India holds second position in production of cardamom in the world after Guatemala. In the recent past, there is much fluctuation in the cardamom prices, and it is labor-intensive crop so there is decline in the cardamom area and production. In the world after saffron and vanilla cardamom is third most expensive spice. Cardamom is valued for its flavor and pungency. α-terpinyl acetate and 1,8-cineole are the two important active ingredients which produce characteristic aroma in cardamom. It is one of the important ingredients in many Indian cuisines. It finds place in ancient as well as modern medicines due to its immense medicinal properties. By seeing all the benefits and uses of cardamom, thrust has to be given for increasing the production and productivity of cardamom which ultimately helps India to become world leader in cardamom production.

Keywords

Queen of spices · Oleoresin · Essential oil · Medicine · Pungency · Flavor

1.1 Introduction

Cardamom is the most expensive spice crop in the world after saffron and vanilla, and it is rightly referred as "queen of spices" due to its characteristic aroma and flavor. It belongs to the genus *Elettaria* and species *cardamomum*. Cardamom belongs to zingiberaceae family and is an herbaceous rhizomatous plant originated from evergreen forests of Western Ghats. It is cultivated in an elevation of 700–1500 m above MSL. Dried fruits of the cardamom plant are the cardamom of commerce.

It is one of the earliest known and extensively used spice crop in the world. Cardamom is one of the important ingredients of many cuisines. It finds an important place in food, flavoring, perfumery, medicinal, and beverage industry. Seeds of cardamom have vast medicinal uses; hence, it is one of the components in many medicines since time immemorial.

H. J. Akshitha (✉) · M. S. Shivakumar ·
S. J. Ankegowda · P. Mohammed Faisal · H. Asangi ·
M. Balaji Rajkumar
ICAR-Indian Institute of Spices Research,
Regional Station, Appangala, Madikeri, Kodagu (D),
Karnataka 571201, India
e-mail: akshi.hosahalli10@gmail.com

S. Aravind · D. Prasath
ICAR-Indian Institute of Spices Research,
Kozhikode, Kerala 673012, India

K. K. Sabu and C. Kole (eds.), *The Cardamom Genome*, Compendium of Plant Genomes,
https://doi.org/10.1007/978-3-031-71829-8_1

Cardamom is grown in India, Guatemala, Sri Lanka, Thailand, Malaysia, Indonesia, El Salvador, Nepal, Tanzania, Vietnam, Costa Rica, and Honduras. Largest producer of cardamom is Guatemala in the world followed by India.

1.1.1 History

Cardamom cultivation is known from ancient time and in Sanskrit is mentioned as *ela* (Cumo 2013). Health benefits of this spice were recognized by Babylonians and Assyrians (Parry 1969). Greek physicians Dioscorides and Hippocrates mentioned cardamom as digestive aid. There was huge demand for cardamom in ancient Greece and Rome, and its trade was luxury business (Cumo 2013). In 176 AD, cardamom was eligible for import tax among the various spices in Alexandria. There was a mention of lesser and greater cardamom which are used in South India in the Journal of Indian Travels by Linschoten (Watt 1872).

Until the discovery of sea route and Portuguese landed in the West Coast of India, cardamom was regarded as spice produced in Arabia. This leads to the end of Arabian monopoly in spices trade. Later Portuguese started collecting and exporting of cardamom, pepper, and ginger from Europe. Among these three spices, European colonizers were more interested in procurement of black pepper and ginger during sixteenth–eighteenth century. During that time, cardamom was regarded as minor forest produce. Later during nineteenth century cardamom cultivation began. Cardamom was under cultivation as secondary crop in coffee plantations. Whereas in the hilly terrains of Western Ghats and the South of Palghat cardamom cultivation spread rapidly. Those areas were known as cardamom hills (Ravindran 2002).

Earlier in Cochin and Travancore, the respective Governments were having monopoly over cardamom. In Travancore, the *Raja* wanted all the cardamom produce should be sold to his officials from them; it will be moved to Alleppey, a port in Travancore. The cardamom sold through auction in Alleppey was majorly purchased by *Moplah* merchants, and then, it was transported to different parts of India and also to England through third parties. Only the finer quality cardamom was shipped to Europe. In Coorg forest, lands were leased for cultivation of cardamom (Watt 1872).

Bleaching cardamom was practiced in the earlier days also, and good market was there for bleached cardamom in Bombay and Bangalore. There was no mention of any chemicals for the purpose of bleaching. To produce bleached cardamom, the produce from Dakshina Kannada was taken to Dharwad (Haveri). Cardamom taken to Dharwad was bleached with the use of water from a well which helps in bleaching the produce, and flavor of cardamom was also improved. Besides bleaching at Sirsi cardamoms were also starched. The starched cardamoms were looking whiter than the bleached one. Nee, wheat, soap, and butter milk were used in preparation of starch. This starch was dissolved in water and sprinkled over the cardamom capsules which were then rubbed with hand (Watt 1872).

1.1.2 Botany

Many species under Zingiberaceae are important ornamental plants, spices, and medicinal plants. Small cardamom, large cardamom, turmeric, and ginger are the commercial Zingiberaceous spices traded (Singh et al. 2012). Genus *Elettaria* consists of 8 species and is known from Sri Lanka to Malaysia and Indonesia. Cardamom plant stature is characterized by leafy shoots, and the inflorescences are born from the rhizome. Flowers in cincini with tubular bracteoles. Cardamom of commerce is dried capsules of *Eletttaria cardamomum* Maton, and in India, it is the monotypic species. It is originated from the moist evergreen forest in South India especially Southern-Western Ghats.

Gregory (1936) reported basic chromosome number of cardamom as $x = 12$ and $2n = 48$. Cardamom is balanced tetraploid.

Leafy shoots are up to 4 m height, and petioles are 2.5 cm long, lanceolate, and acuminate. Leaves are lightly pubescent or glabrous below.

Inflorescence are characterized as prostrate or erect or semi-erect based on the growth habit, and these inflorescence arise from bottom of the shoot. Many flowers are found in each cincinni. Bracteoles are 2.5 cm long, tubular, mucronate, and glabrous. Calyx up to 2 cm long, 2- or obscurely 3-lobed, and lobes mucronate. Corolla tube about as long as calyx, lobes 1–1.5 cm long, rounded at the apex, the dorsal tube widest. Labellum is white with violet streaks. Lateral staminodes inconspicuous, subulate. Anther sessile, 2–3 mm long ovary, glabrous. Fruits are botanically called as capsule and are oblong or more or less globose in shape (Burtt and Smith 1983).

1.1.3 Ecotypes

Based on the nature of panicles, cardamom is characterized into three ecotypes, viz. Mysore, Malabar, and *Vazhukka* (Sudharshan et al. 1991; Madhusoodanan et al. 2002).

1.1.3.1 Malabar

Malabar type plants are dwarf (2–3 m height). Leaf dorsal side will be pubescent or glabrous. Leaf petiole will be short, and panicles are prostrate. Fruits are globose to oblong-shaped. Malabar type of plants can withstand long dry spell of up to 4–6 months. These are suitable for cultivation in lower elevation (600–100 m MSL). Capsules will be pale/golden yellow in color.

1.1.3.2 Mysore

Mysore type plants will be tall (3–5 m height) and robust. Leaves are oblong-lanceolate, lanceolate, or glabrous on both sides. Leaf petiole will be long, and panicles are erect. This type needs well distributed rain as well as assured irrigation and are suitable for cultivation in higher elevation (900–1200 m MSL). Capsules will be ovoid, bold, and are green in color.

1.1.3.3 Vazhukka

Vazhukka type is considered as natural hybrid between Malabar and Mysore type. In this type, plants are robust and grow to a height of 3–5 m. Leaves will be deep green, oblong-lanceolate or ovate. As like Mysore type, this type also needs well-distributed rain as well as assured irrigation and are suitable for cultivation in higher elevation (900–1200 m MSL). Panicle will be semi-erect as an intermediate between Malabar and Mysore. Capsules are bold, globose, or ovoid in shape, and capsules will be dark green in color.

1.2 Area and Production

India is the global leader in the production, consumption, and export of spices. Until early eighties India had monopoly in area, production, and export of cardamom, and thereafter, Guatemala overtook India and became leader in production and productivity. Cardamom is grown in only a limited unit of the Western Ghats of India. Kerala, Tamil Nadu, and Karnataka are the primary growing regions for cardamom in India. Kerala holds the most shares of the total production followed by Karnataka and Tamil Nadu. Based on data, the cardamom-growing region in India has not changed significantly over the past five years. The production of cardamom during 2022–23 was 24,463 tons from an area of 70,410 ha.

In Kerala, cardamom is cultivated in nine districts, and Idukki stands first in area and production. In Karnataka, Kodagu, Chikkamagaluru, Hassan, and Dakshina Kannada are the major cardamom-producing districts. In Tamil Nadu, cardamom is cultivated in Madurai, Salem, Coimbatore, Ramnathapuram, Nilgiris, and Tirunelveli (Fig. 1.1; Table 1.1).

In India, only 15–20% of total production is exported, with the remaining 80–90% goes for domestic consumption. India is the second-largest exporter after Guatemala. Indian cardamom is favored on the international market because of its superior quality and aroma. Cardamom oil and oleoresin, two value-added goods, are exported to European nations. The COVID-19 pandemic did not stop India's spice exports from growing. In 2022–23, 7352.37 MT of cardamom worth Rs. 87,514.84 Lakhs was exported. Saudi

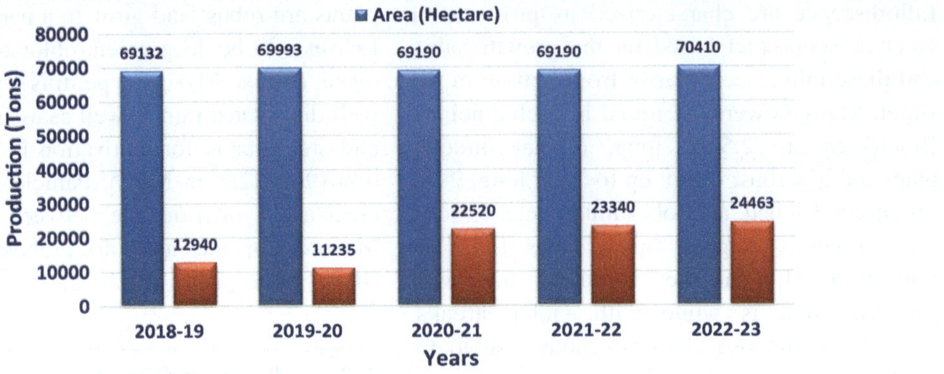

Fig. 1.1 Total area and production of cardamom in India during last five years

Table 1.1 State-wise area and production of cardamom in India

(Area in hectare, production in tons)										
State	2018–19		2019–20		2020–21		2021–22		2022–23	
	Area	Production	Area	Production	Area	Production	Area	Production	Area	Production
Kerala	38,882	11,535	39,697	10,075	39,143	20,570	39,143	21,270	40,345	22,165
Karnataka	25,135	690	25,135	620	25,135	579	25,135	697	25,135	833
Tamil Nadu	5115	715	5162	540	3892	1372	4912	1373	4930	1466
Total	69,132	12,940	69,993	11,235	69,190	22,520	69,190	23,340	70,410	24,463

Source Spices Board

Table 1.2 Export of small cardamom from India

S. No.	Years	Quantity (MT)	Value (Lakhs)
1	2018–19	2849.96	35,625.52
2	2019–20	1862.85	42,604.79
3	2020–21	6486.53	110,346.57
4	2021–22	10,571.25	137,566.98
5	2022–23	7352.37	87,514.84

Source Spices Board

Arabia, Japan, Malaysia, United Kingdom, and Kuwait are the major importers of cardamom and cardamom value-added products from India. Largest importers of Indian cardamom are Saudi Arabia and UAE (Table 1.2).

1.2.1 Global Scenario

According to the Trade Promotion Council of India, the cardamom sector is predicted to increase at a compound annual growth rate (CAGR) of 2.8% from its estimated value of US $69.8 billion in 2018 to US $86.4 billion by 2025. The cardamom market is extremely consolidated worldwide, with Guatemala and India accounting for 60% of production.

Guatemala is the world's largest cardamom producer. Other important cardamom-producing nations are India, Sri Lanka, Nepal, and Indonesia. An estimated 35,000 MT of cardamom is produced worldwide each year, of which 25–30,000 MT are produced in Guatemala. The consumption of cardamom has increase globally over the past 20 years. Middle Eastern nations, India, Pakistan, European nations, the United States, and Japan are the main consumers of cardamom. Saudi Arabia and the United Arab Emirates consume more than 60% of the cardamom used worldwide. Saudi Arabia holds a market share of approximately 60% for Indian cardamom, with Kuwait, the United Arab Emirates, and United States.

1.3 Chemical Composition, Constituents Responsible for Flavor, and Pungency

Cardamom's distinctive aroma and flavor come from its essential oil, and the specific makeup of this oil's volatile components creates its unique scent and taste. The inherent sensory properties of cardamom are not influenced by the color of its fruits. Freshly harvested cardamom capsules contain more oil than those that have been stored for extended periods. The highest volatile oil content occurs 20–25 days before full maturity, with longer durations leading to increased evaporation. Cardamom seeds have a single layer of cells just below the epidermis that contains the essential oil, and fine grinding can cause a loss of these volatile oils (Zachariah 2002).

"Cardamom essential oil" is the aromatic compound found in cardamom, typically extracted through cold pressing and water or steam distillation methods. It is often described as having a spicy, sweet, warm, slightly camphor-like, and citrusy scent (Robert 1986; Boiswert and Hubert 1998). Chemical composition is influenced by genotype/variety, environmental condition in which it is grown/region, drying procedures, storage conditions, and storage duration. Seed volatile oil ranges from 2 to 5% and is influenced by storage condition and storage duration. 1,8-cineole and α-terpinyl acetate are the two key components responsible for the distinctive aroma of cardamom (Lawrence 1978; Olivero-Verbel et al. 2010). 1,8-cineole is the compound that imparts a camphor-like aroma, with higher concentrations found in the Malabar type of cardamom. On the other hand, the Mysore type has a milder spicy flavor due to its greater content of α-terpinyl acetate. 1,8-cineole, an oxygenated monoterpene, is also found in various other plant essential oils, such as eucalyptus (Rodríguez et al. 2006) and bay leaves (Sangun et al. 2007), where it is believed to contribute to their characteristic aroma. Cardamom oil has a warm, slightly pungent, aromatic, and camphoraceous flavor, making it popular as a spice, a flavoring agent in tea, food preparations, perfumery, medicines, and beverages (Chempakam and Sindhu 2008). A significant amount of cardamom is used in pan masalas and for chewing. 1,8-cineole is the primary compound in cardamom, commonly used as a food flavoring agent (Santos and Rao 2001).

Yield of essential oil in cardamom ranges from 0.2 to 5.6 (Subaddarage et al. 1985), 2.6 (Sontakke et al. 2018), 1.0 (Ebru and Zehra 2013), 6–14 (Menon 2000), 4–7 (Abbasipour et al. 2011), and 6–8.7 (Leela et al. 2008); first two workers used steam distillation and later 4 used hydro-distillation. Region in which cardamom cultivated also plays an important role in composition of constituents. Cardamom cultivated in India contains α-terpinyl acetate, 1,8-cineole, and α-terpineol as major components (Sharma et al. 2011; Savan and Kucukbay 2013), whereas α-terpinyl acetate, 1,8-cineole, sabinene, linalyl acetate, and linalool are key constituents in Guatemala cardamom seeds (Singh et al. 2008). 1,8-cineole (25.6%), linalool (6.3%), and α-terpinyl acetate (40.7%) are key constituents in Turkey cardamom (Gochev et al. 2012). Thomas et al. (2006) analyzed the cardamom samples from India, Sri Lanka, and Guatemala and reported that quality-wise Indian cardamom was superior over Sri Lankan and Guatemalan cardamom. Major aroma and flavor imparting components like α-terpinyl acetate and 1,8-cineole were high in Indian cardamom, thus strengthening the credence of the high intrinsic quality of the Indian cardamom. Pillai et al. (1984) found that Guatemalan cardamom oil is slightly superior to Indian cardamom oil due to its higher content of α-terpinyl acetate. In a study by Noleau et al. (1987), the volatile composition of Costa Rican cardamom was found to be similar to that of other cardamom-producing regions. Environmental factors such as shade can also affect the quality of cardamom. For example, full sunlight conditions lead to an increase in the sweet aroma due to higher levels of α-terpinyl acetate, while 75% shade enhances the camphorous flavor by increasing the concentration of 1,8-cineole (Alagupalamuthirsolai et al. 2019).

The primary components of cardamom essential oil are 1,8-cineole (ranging from 20 to 60%) and α-terpinyl acetate (20–55%) (Padmakumari et al. 2010). According to Ashokkumar et al. (2019a, b, 2021), major constituents of cardamom essential oil include 1,8-cineole (28.94%), α-terpinyl acetate (26.7%), α-terpineol (14.6%), sabinene (13.5%), nerol (5.0%), and α-pinene (2.4%). Mejdi et al. (2015) noted additional minor constituents such as limonene (2.9%), 4-terpineol (1.4%), α-pinene (1.1%), β-pinene (0.8%), myrcene (0.8%), octanal (0.2%), δ-3-carene (0.4%), p-cymene (0.7%), (E)-nerolidol (0.7%), cis-sabinene hydrate (0.6%), geranyl acetate (0.3%), cis-sabinene hydrate acetate (0.2%), β-caryophyllene (0.2%), β-selinene (0.2%), γ-cadinene (0.2%), and trans-linalool oxide (0.1%). Similar composition of CEO was reported by Lawrence (1978), Govindarajan et al. (1982), Noleau et al. (1987), Korikanthimath et al. (1997), Miniraj et al. (2000), Murugan et al. (2002, 2005), Kumar et al. (2005), Kaskoos et al. (2006), Thomas et al. (2006), Chegini and Abbasipour (2017), Alagupalamuthirsolai et al. (2019, 2020), Ashokkumar et al. (2019a, b), and Alam et al. (2019). Not only seed even some reports indicate the presence of essential oil and its components in cardamom leaves. Major constituents of leaf essential oil are α-terpinolene, p-cymene, α-terpinene and α-thujene (Mahmud 2008), terpinen-4-ol (Sharma et al. 2020), and 1,8-cineole (Jena et al. 2021). GC–MS analysis of leaf and stem led to identification of 43 and 37 compounds representing 92.71% and 92.12% of the total leaf and stem oil, respectively (Jena et al. 2021). Many spice oils contain alcohols, esters, and aldehydes, but what makes cardamom oil unique is its high concentration of the ether 1,8-cineole and the esters α-terpinyl acetate and linalyl acetate. α-terpinyl contributes to the bitterness in cardamom oil, typically ranging from 0.8 to 2.7% (Raghavan et al. 1991). Additionally, flavor quality is influenced by other components like borneol and citral.

Zachariah (2002) remarked that cardamom oil is a remarkable gift from nature, crafted through a complex array of terpenes, terpene alcohols, esters, and other compounds that challenge even the most advanced analytical methods. Creating "synthetic cardamom oil" with the same sensory qualities from these components is still beyond human capabilities today (Table 1.3).

Oleoresin is derived through solvent extraction and typically represents the flavor quality of cardamom more accurately than the volatile oil. In addition to aromatic compounds, oleoresins contain non-aromatic fats, waxes, resins, colors, and other substances that dissolve in the chosen solvent. Recently, supercritical fluid extraction (SCFE) has also been used to obtain oleoresins. However, oleoresin is typically produced on a relatively small scale and is not used directly in perfumery unless it is refined to remove the fat components, yielding a more purified product known as concrete.

1.4 Cardamom for Food

Cardamom finds a unique place in kitchens worldwide, featured in both traditional and modern culinary endeavors. This versatile spice serves as a flavor enhancer in a wide array of dishes, ranging from savory to sweet. Whether it is infused in teas, coffees, snacks, desserts, or even alcoholic beverages, cardamom seamlessly integrates into diverse cuisines. It is available in three primary forms: whole pods, decorticated seeds, and ground powder. While globally traded in its whole pod form, the market for decorticated seeds is limited, and powdered cardamom constitutes a negligible portion. Decorticated seeds, though offering a lower price point, suffer from the loss of volatile oils during transportation and storage, affecting their market value (Korikanthimath et al. 2006).

Cardamom serves as a quintessential ingredient in domestic kitchens worldwide, whether in its whole or ground form. Across Asian nations, it holds a vital place in spiced rice dishes, vegetable and meat preparations, as well as flavoring soft drinks, payasa, and ghee rice. In Arab cultures, the esteemed beverage known as Gahwa features prominently, with cardamom-infused coffee being a symbol of prestige. In

Table 1.3 Constituents responsible for flavor in cardamom

Components	Flavor description	Effect in flavor use	Concentration range in cardamom oil (%)
Esters			
α-terpinyl acetate	Aroma is mildly herbaceous sweet-spicy blend with hints of pine, and the flavor offers a warm, slightly spicy taste	To stretch cardamom herbal variation spice, imitation citrus, and cherry peach flavors	34–52
Linalyl acetate	Sweet, floral, and fruity aroma and taste, with lower tenacity but more intensity than terpinyl acetate	It is a fresh, sweet enhancer used in perfumes and berry-flavored products	0.7–6.3
Geranyl acetate	Sweet, floral, fruity scent with a hint of green, offering a stronger note than geraniol	Sweetener, modifier in citrus fragrances and fruit, citrus, and spice flavors	0.17–0.23
Neryl acetate	Very sweet; fruity floral	Effective in berry flavors, higher levels in perfumery	0.17–0.23
Ethers			
1,8-cineole	Fresh, camphoraceous cool and odor and taste with high diffusivity but low tenacity	It has a refreshing effect, making it widely used in perfumes and flavoring applications	23–51
Alcohols			
Linalool	It has a floral, woody aroma with a note of citrus, and a creamy floral taste when used in small quantity	It imparts a uniquely pleasant taste when used in small amounts	1.4–4.5
α-terpineol	Delicately floral, sweet, and reminiscent of lilacs	Citrus and spice blends or formulations	1.4–3.3
Citronellol	Fresh rosy odor and floral rosy bitter taste	Extensive use in perfumes and fruit flavors	0.01–0.04
Geraniol	Floral, rosy with warm dry tones	Berry and sweet-spicy flavor	0.29–0.64
Nerol	Sweet rosy, fruity	Berry and sweet-spicy flavors	0.29–0.64
4-terpineol	Warm and peppery with woody undertones, complemented by earthy, musty notes, and a pleasantly green aroma	Primarily used in citrus and spice blends, offering warm and herbaceous effects	0.14–0.87
Nerolidol	Woody, floral slightly green	Excellent tenacity	0.28–1.6
Carbonyls			
Camphor	It has a warm, minty odor, with a bitter, warm taste followed by a cool mouthfeel	Modifier diffusive	0.02–0.04
Citral	Powerful lemon fruity odor	Commonly used in high dilutions for citrus and spice blends	0.19–0.26

Source Zachariah (2002)

North America and Europe, ground cardamom is commonly utilized in soups, curry blends, sausage recipes, and canned fish. The versatility of cardamom extends to innovative culinary creations such as cardamom cola, biscuits, toffees, and pastries, as well as modern adaptations like carbonated Gahwa and encapsulated cardamom oil. Its usage spans across various culinary traditions: from Arab coffee culture to German curry blends, American baking, Russian confectionery, Japanese savory dishes, and Scandinavian sweets. Notably, Indian cardamom stands out for its nutritional value, being low in fat yet rich in vitamins, iron, and protein.

Oils and oleoresins derived from cardamom find versatile applications in various industries, including flavoring cordials, cosmetics, perfumery, chewing gums, and medicinal formulations. Known for its stimulating properties, cardamom holds a significant place in traditional medicine. Cardamom tea is often consumed to alleviate conditions such as scanty urination, exhaustion from overwork, and dysentery. Additionally, when combined with honey, cardamom is believed to enhance eyesight and fortify the nervous system. As a key component of digestive stimulants, it aids in relieving flatulence and indigestion. Gargling with a concoction of cinnamon and cardamom infusion is said to treat pharyngitis, hoarseness, sore throat, and offering relief during flu infections. Regular gargling with this mixture is believed to provide protection against flu viruses. Moreover, incorporating powdered cardamom seeds into tea not only imparts a delightful aroma but also offers potential medicinal benefits, promoting overall health. For smokers aiming to reduce cigarette consumption, consuming a few cardamom seeds daily may serve as a preliminary step toward minimizing dependency on chain smoking (Korikanthimath 2000).

Commercial production of cardamom oil involves steam distillation of crushed fruits or powdered seeds. Capsules harvested during physiologically mature and immature stages exhibit better color retention and higher oil recovery (20–30%) compared to fully ripe ones (Korikanthimath 1993). This finding was corroborated by Leela et al. (2008), who recommended harvesting capsules at 110–124 days after flowering (physiologically mature stage) for optimal oil yield. However, despite its high oil content, cardamom essential oil is susceptible to developing off-flavors upon exposure to air over time. Consequently, its usage is typically confined to fresh meat products and foods with short shelf lives (Korikanthimath 2000).

1.5 Health Benefits of Cardamom

a. Use of Cardamom in Ancient Medicine

The use of cardamom in ancient medicine is well documented by German Egyptologist Georg Ebers in a medical document dated 1550 BC, and the Egyptians used cardamom in medicine, cosmetic ointments, aromatic oils, cooling, ointments, fumigation, and embalming (Rosengarten 1969). Blend of spices including small cardamom were included in the ancient Susruta *mushkakadigana* (herbal medicine) which were prescribed to remove excess fatty storage and also to cure urinary complaints, jaundice, and piles. In traditional medicine, cardamom is used as an integrated part for treating aliment such as disorders affecting kidney and urinary system (Ballabh et al. 2008) and also works as protective agents in gastrointestinal disorders (Jafri et al. 2001). Early Egyptians used blends of cardamom as aromatic body ointment and pomades. Spices inclusive of ginger, black pepper, small cardamom, mustard seed, and cumin were part of primeval Susruta *mushkakadigana* (herbal remedies), possessing the properties to remove excess fat and to cure complex ailment of urinary bladders, piles, and liver diseases. The records for uses of cardamom were documented in earliest Brahminic texts, and to increase the flow of saliva and help in digestion, betel leaves were used to encase a blend of cloves and cardamom for chewing (Rosengarten 1969).

Ever since fourth century BC, cardamom has been used in ayurvedic medicine (respiratory issues like bronchitis and asthma, as well as constipation) (Al-Zuhair et al. 1996; Bisht et al. 2011; Korikanthimath et al. 2000), treating food poison (Ashokkumar et al. 2019a, b) and cooking. The medicinal properties were also mentioned in Egyptian papyrus during 1550 BC, and in addition ancient Greeks and Romans also utilized into their cuisine and medical treatments (Korikanthimath et al. 2000). Uses of cardamom are also reported in Chinese traditional medicine as a remedial measure for digestion-related ailments and bladder infection (Kapoor 1990; Duke et al. 2003). Susruta II has described over seven hundred drugs of plant origin including cardamom as one of those herbs in cure of several diseases. Honey mixed with cardamom powder improves the eyesight, and it used in traditional medicine (Singh and Singh 1996). In traditional ayurvedic medicines, several important preparations from cardamom capsules as powders, decoctions, oils, and beverages such as *Arishta* and *Aasava* are used for curing several diseases (Sahadevan 1965). Cardamom is an important spice which is believed to have tridoshic property balancing all the three doshas—"*Vatta, Pitta, and Kapha*" (Rani et al. 2018).

b. **Cardamom in Modern Medicine**

Due to advancement in ayurvedic and allopathic medicines, the medicinal use of cardamom has increased to a huge extent in particular cardamom oil, which has a high market value. Cardamom is a good candidate for volatile oils, phenolics acids, essential oils, oleoresins, and sterols (Naveed et al. 2013). Cardamom oil is extensively utilized in the production of certain non-chemical hand creams and soaps on a large scale (Ajmera et al. 2018). In addition to this, the volatile oils are found to exhibit anti-inflammatory, antispasmodic, analgesic, anti-obesity, and numerous antimicrobial principles (Majdalawieh and Carr 2010; Aggarwal et al. 2011). Other medicinal uses reported in cardamom are treatment ailments of cardiovascular, renal, debility, bronchitis, asthma, and gastrointestinal disorders (Sengupta et al. 2009). Supplementation of cardamom powder prevents the dyslipidaemia and hepatic damage, and it alleviates oxidative stress in rats on a high carbohydrate and high-fat diet and helps reduce obesity (Rahman et al. 2017).

Beverages crafted with powdered cardamom serve as an antidote against venom from snakes and scorpions, and it also helps in reduction of inflammation and headaches (Govil 1998). "Elakkai tea", a pleasant aromatic tea prepared by crushing cardamom capsules and boiled with tea and water, reduces the tiredness and depression, and this is due to the presence of β-carotene ($0.5\ \mu g\ g^{-1}$) (Ashokkumar et al. 2019a, b). Cardamom is predominantly used in oral health in order to avoid bad breath, anti caries, antiseptic, and antimicrobial (Jesylne et al. 2016). Presently cardamom is used in modern medicine to combat metabolical dysfunction and prevent free radicals, oral bacteria, and bad breath, and the compound cineole found in cardamom reported to be antiseptic and arrest bacterial growth (Aneja and Joshi 2009), fights bacterial and fungal infection (Agnihotri and Wakode 2010), and reduces muscle and joint pain (Aggarwal et al. 2011; Al-Zuhair et al. 1996). The cytotoxic ability of natural killer cells increases upon treatment with cardamom extracts showing a possible anticancer property (Qiblawi et al. 2020).

c. **Different Medicinal Properties of Cardamom**

Certain compounds found in cardamom, such as 1,8-cineole, α-terpinyl acetate, α-terpineol, sabinene, and nerol, as identified by Ashokkumar et al. in 2019, contribute to its medicinal properties. Additionally, research by Chegini and Abbasipour (2017), Singh et al. (2008), and Noleau et al. (1987) highlights the presence of linalyl acetate, linalool, and geraniol. These compounds are known for their antioxidant, anti-inflammatory, and antimicrobial properties, which are beneficial for various health conditions. Their presence in cardamom underscores its potential as a natural remedy for ailments and emphasizes its role beyond just a culinary spice. In addition, several flavonoids and carotenoids have huge potential in the medicine (Ashokkumar et al. 2019a, b). Furthermore, Parry et al. (2006) have identified several fatty acids present in cardamom, including myristic acid, palmitic acid, palmitoleic acid, stearic acid, oleic acid, linoleic acid, α-linolenic acid, arachidic acid, and eicosenoic acid. These fatty acids contribute to the health benefits associated with cardamom consumption. They play vital roles in various physiological processes, such as maintaining cardiovascular health, supporting brain function, and reducing inflammation. The presence of these fatty acids further underscores the nutritional value and potential therapeutic properties of cardamom.

Cardamom possess following medicinal properties

i. **Antimicrobial**

Cardamom is predominantly utilized as an antimicrobial agent, as indicated by studies conducted by Gradinaru et al. (2014). It is commonly employed to treat infections in teeth and gums due to its antimicrobial properties. The aqueous extract of cardamom capsule has also been used against *Candida* spp. (yeast) and *Saccharomyces cerevisiae* (Noumi et al. 2018).

ii. **Antidote to Snake Venom**

Cardamom is employed as an antidote for snake and scorpion venom, drawing upon its medicinal attributes to mitigate the effects of envenomation. This traditional practice, rooted in ancient wisdom, offers a natural remedy in regions prone to encounters with venomous creatures (Rajathi et al. 2017).

iii. **Cardiovascular Protection**

Cardamom extracts were found to reduce arterial pressure both systolic and diastolic and also certain myocardial enzymes (Goyal et al. 2015), thus preventing the chances of myocardial infarction (heart attack).

iv. **Antidiabetic**

Patients with different stages of diabetic such as pre-diabetic and Type II diabetic showed little variation in glycemic ratio indicating the possible role of cardamom in reducing the blood sugar level (Winarsi et al. 2014).

v. **Hepatoprotective**

The compounds present in volatile oils such as 1,8-cineole, sabinine, and α-pinene act as stimulant for liver which promotes the elimination of bile and reduces liver congestion (Rajathi et al. 2017). It also helps in reducing the tumor in non-alcoholic fatty liver (Daneshi et al. 2018).

vi. **Antiulcerogenic**

A protective effect from ethanolic extract of cardamom has been found in rats lesions induced by aspirin treatment by inhibiting the gastric mucosal barrier (Jamal et al. 2006).

In addition to this, cardamom also possess several other properties such as anticancer, anti-inflammatory, anthelminthic, diuretic, aphrodisiac, antiseptic, and stomachic principles.

References

Abbasipour H, Mahmoudvand M, Rastegar F et al (2011) Fumigant toxicity and oviposition deterrency of the essential oil from cardamom, *Elettaria cardamomum* against three stored-product insects. J Insect Sci 11:165. https://doi.org/10.1673/031.011.16501

Aggarwal BB, Prasad S, Reuter S et al (2011) Identification of novel anti-inflammatory agents from ayurvedic medicine for prevention of chronic diseases: "reverse pharmacology" and "bedside to bench" approach. Curr Drug Targets 12(11):1595–1653. https://doi.org/10.2174/138945011798109464

Agnihotri S, Wakode S (2010) Antimicrobial activity of essential oil and various extracts of fruits of greater cardamom. Indian J Pharm Sci 72(5):657–659

Ajmera P, Singh A, Chauhan et al (2018) Cardamom crop production and harvesting: a review. Int J Food Sci Nutr 3:174–178

Alagupalamuthirsolai M, Ankegowda SJ, Murugan M et al (2019) Influence of light intensity on photosynthesis, capsule yield, essential oil and insect pest incidence of small cardamom (*Elettaria cardamomum* (L.) Maton). J Essent Oil Bear Plants 22:1172–1181. https://doi.org/10.1080/0972060X.2019.1690587

Alagupalamuthirsolai M, Sivaranjani R, Ankegowda SJ et al (2020) A study of essential oil constituents from capsules, physiological and quality parameters of small cardamom ecotypes [*Elettaria cardamomum* (L.) Maton] growing in the Western Ghats, India. J Essent Oil Bear Plants 23(6):1253–1264. https://doi.org/10.1080/0972060X.2020.1869105

Alam A, Majumdar RS, Alam P (2019) Systematics evaluations of morphological traits, chemical composition, and antimicrobial properties of selected varieties of *Elettaria cardamomum* (L.) Maton. Nat Prod Commun 14(2):1–7. https://doi.org/10.1177/1934578X19892688

Al-Zuhair H, El-Sayeh B, Ameen HA et al (1996) Pharmacological studies of cardamom oil in animals. Pharmacol Res 34(1–2):79–82

Aneja KR, Joshi R (2009) Antimicrobial activity of *Amomum subulatum* and *Elettaria cardamomum* against dental caries causing microorganisms. Ethnobot Leaflets. https://opensiuc.lib.siu.edu/ebl/vol2009/iss7/3

Ashokkumar K, Murugan M, Dhanya MK et al (2019a) Botany, traditional uses, phytochemistry and biological activities of cardamom [*Elettaria cardamomum* (L.) Maton]—a critical review. J Ethnopharmacol. https://doi.org/10.1016/j.jep.2019.112244

Ashokkumar K, Murugan M, Dhanya MK et al (2019b) Phytochemical variations among four distinct varieties of Indian cardamom *Elettaria cardamomum* (L.) Maton. Nat Prod Res 34(13):1919–1922. https://doi.org/10.1080/14786419.2018.1561687

Ashokkumar K, Vellaikumar S, Murugan M et al (2021) Essential oil profile diversity in cardamom accessions from southern India. Front Sustain Food Syst 5:639619. https://doi.org/10.3389/fsufs.2021.639619

Ballabh B, Chaurasia OP, Ahmed Z et al (2008) Traditional medicinal plants of cold desert Ladakh—used against kidney and urinary disorders. J Ethnopharmacol 118(2):331–339

Bisht VK, Negi JS, Bhandari AK et al (2011) *Amomum subulatum* Roxb: traditional, phytochemical and biological activities—an overview. Afr J Agric Res 6:5386–5390

Boiswert C, Hubert A (eds) (1998) L'ABCdaire des épices. Flammarion, Paris

Burtt BL, Smith RM (1983) In: Dassanayake MD (ed) Zingiberaceae. A revised hand book to the flora of Ceylon, vol IV. Amerind Pub., New Delhi

Chegini SG, Abbasipour H (2017) Chemical composition and insecticidal effects of the essential oil of cardamom *Elettaria cardamomum*on the tomato leaf miner *Tuta absoluta*. Toxin Rev 36:12–17

Chempakam B, Sindhu S (2008) Large cardamom. In: Parthasarathy VA, Chempakam B, Zachariah TJ (eds) Chemistry of spices. CAB International, pp 59–69. https://doi.org/10.1079/9781845934057.0059

Cumo CM (2013) Encyclopedia of cultivated plants: from acacia to zinnia, 3 vols. ABC-CLIO. ISBN 978-1-59884-775-8

Daneshi MM, Keshavarz SA, Qorbani M et al (2018) Green cardamom increases Sirtuin-1 and reduces inflammation in overweight or obese patients with non-alcoholic fatty liver disease: a double-blind randomized placebo-controlled clinical trial. Nutr Metab 15:63. https://doi.org/10.1186/s12986-018-0297-4

Duke JA, Bogenschutz-Godwin MJ, de Cellier J et al (2003) *Elettaria cardamomum* (L.) Maton (Zingiberaceae) Cardamom, Malabar or Mysore cardamom. In: CRC handbook of medicinal spices. CRC Press, Washington, DC, p 159

Ebru KS, Zehra FK (2013) Essential oil composition of *Elettaria cardamomum* Maton. J Appl Biol Sci 7:42–45

Gochev V, Girova T, Stoilova I et al (2012) Low temperature extraction of essential oil bearing plants by liquefied gases. 7. Seeds from cardamom (*Elettaria cardamomum* (L.) Maton). J Biosci Biotech 1(2):135–139

Govil JN (1998) Glimpses in plant research, vol XII. Current concepts of multidiscipline approach to the medicinal plants (part I). Today and Tomorrow's Printers, New Delhi

Govindarajan VS, Shanti Narasimhan, Raghuveer KG et al (1982) Cardamom—production, technology chemistry and quality. CRC Crit Rev Food Sci Nutr 16:326

Goyal SN, Sharma C, Mahajan UB et al (2015) Protective effects on cardamom in isoproterenol induced myocardial infraction in rats. Int J Mol Sci 16:27457–27469

Gradinaru AC, Aprotosoaie AC, Trifan A et al (2014) Interactions between cardamom essential oil and conventional antibiotics against *Staphylococcus aureus* clinical isolates. Farm 62:1214–1222

Gregory PJ (1936) The floral morphology and cytology of *Elettaria cardamomum* Maton. J Linn Soc 50:363–391

Jafri MA, Farah JK et al (2001) Evaluation of the gastric antiulcerogenic effect of large cardamom fruits of *Amomum subulatum* Roxb. J Ethnopharmacol 75(2–3):89–94

Jamal A, Kalim J, Aslama M et al (2006) Gastroprotective effect of cardamom *Elettaria cardamomum* Maton. fruits in rats. J Ethnopharmacol 103:149–153

Jena S, Ray A, Sahoo A et al (2021) Chemical composition and antioxidant activities of essential oil from leaf and stem of *Elettaria cardamomum* from Eastern India. J Essent Oil Bear Plants 24(3):1–9. https://doi.org/10.1080/0972060X.2021.1937335

Jesylne P, Soundarajan S, Murthykumar K et al (2016) The role of cardamom oil in oral health: a short review. Res J Pharm Tech 9(3):272–274

Kapoor LD (1990) Handbook of ayurvedic medicinal plants. CRC Press, Boca Raton, p 172

Kaskoos RA, Ali M, Kapoor R et al (2006) Essential oil composition of the fruits of *Eletteria cardamomum*. J Essent Oil Bear Plants 9:81–84

Korikanthimath VS (1993) Harvesting and on farm processing of cardamom. In: Proceedings of national seminar on post harvest technology of spices. ISS Calicut, pp 62–68

Korikanthimath VS (2000) Cardamom. In: Peter KV (ed) Handbook of herbs and spices. CRC Press, Washington, pp 123–133

Korikanthimath VS, Mulge R, Zachariah TJ (1997) Variation in yield and quality characters of cardamom clones. J Med Aromat Plants 19(4):1024–1031

Korikanthimath VS, Prasath D, Govardhana R (2000) Medicinal properties of cardamom *Elettaria cardamomum*. J Med Aromat Crops 22/4A & 23/1A:683–685

Korikanthimath VS, Venugopal MN, Sudharshan MR et al (2006) Cardamom. In: Ravindran PN, Nirmal Babu K, Shiva KN, Johny AK (eds) Advances in spices research history and achievements of spices research in India since independence. Agrobios (India), Jodhpur, India, pp 315–364

Kumar A, Tandon S, Ahmad J et al (2005) Essential oil composition of seed and fruit coat of *Elettaria cardamomum* from South India. J Essent Oil Bear Plants 8:204–207. https://doi.org/10.1080/0972060X.2005.10643446

Lawrence BM (1978) Major tropical spices—cardamom (*Elettaria cardamomum*). In: Essential oils, pp 105–155
Leela NK, Prasath D, Venugopal MN (2008) Essential oil composition of selected cardamom genotypes at different maturity levels. Indian J Hortic 65(3):366–369
Madhusoodanan KJ, Pradip Kumar K, Ravindran PN (2002) Botany, crop improvement and biotechnology of cardamom. In: Ravindran PN, Madhusoodhanan KJ (eds) Cardamom—the genus *Elettaria*. Taylor and Francis, pp 11–68
Mahmud S (2008) Composition of essential oil of *Elettaria cardamomum* Maton leaves. Pak J Biol Sci 60:111–114
Majdalawieh AF, Carr RI (2010) *In vitro* investigation of the potential immunomodulatory and anticancer activities of black pepper (*Piper nigrum*) and cardamom (*Elettaria cardamomum*). J Med Food 13:371–381
Mejdi S, Emira N, Ameni D et al (2015) Chemical composition and antimicrobial activities of *Elettaria cardamomum* L. (Maton) essential oil: a high activity against a wide range of food borne and medically important bacteria and fungi. J Chem Biol Phys Sci 6:248–259
Menon NA (2000) Studies on the volatiles of cardamom (*Elletaria cardamomum*). J Food Sci Technol 37:406–408
Miniraj N, Murugan M, Joseph CR (2000) Evaluation of cardamom (*Elettaria cardamomum* Maton) germplasm. J Spices Aromat Crops 9:55–56
Murugan M, Josephrajkumar A, Backiyarani S et al (2002) Essential oil of cardamom (*Elettaria cardamomum* M): effect of infestation by thrips (*Sciothrips cardamom* Ramk.). Ind Perfum 46:321–323
Murugan M, Josephrajkumar A, Sheeba B et al (2005) Essential oil profile of elite small cardamom (*Elettaria cardamomum* M.) accessions and their interaction with thrips (*Sciothrips cardamom* Ramk.) infestation. Ind Perfum 49:219–224
Naveed R, Hussain I, Tawab A et al (2013) Antimicrobial activity of the bioactive components of essential oils from Pakistani spices against salmonella and other multi-drug resistant bacteria. BMC Complement Altern Med 13:265
Noleau I, Toulemonde B, Richard H (1987) Volatile constituents of cardamom (*Elettaria cardamomum*) cultivated in Costa Rica. Flavor Fragrance J 2:123–127
Noumi E, Snoussi M, Alreshidi MM et al (2018) Chemical and biological evaluation of essential oils from cardamom species. Molecules 23:e2818. https://doi.org/10.3390/molecules23112818
Olivero-Verbel J, González-Cervera T, Guette-Fernandez J et al (2010) Chemical composition and antioxidant activity of essential oils isolated from Colombian plants. Braz J Pharmacogn 20:568–574
Padmakumari AKP, Rani MP, Sasidharan I et al (2010) Chemical composition, flavonoid-phenolic contents and radical scavenging activity of four major varieties of cardamom. Int J Biol Med Res 1:20–24
Parry JW (1969) Spices, vol I. Chemical Pub. Co., New York
Parry J, Hao Z, Luther M et al (2006) Characterization of cold-pressed onion, parsley, cardamom, mullein, roasted pumpkin, and milk thistle seed oils. J Am Oil Chem Soc 83:847–854
Pillai OGN, Mathulla T, George KM et al (1984) Studies in cardamom 2. An appraisal of the excellence of Indian cardamom (*Elettaria cardamomum* Maton). Indian Spices 21(2):17–25
Qiblawi S, Kausar MA, Shahid SMA et al (2020) Therapeutic interventions of cardamom in cancer and other human diseases. J Pharm Res Int 32(22):74–84
Raghavan B, Abraham KO, Sankaracharya NB et al (1991) Cardamom—studies on quality of volatile oil and product development. Indian Spices 28(3):20–24
Rahman MM, Alam MN, Ulla A et al (2017) Cardamom powder supplementation prevents obesity, improves glucose intolerance, inflammation and oxidative stress in liver of high carbohydrate high fat diet induced obese rats. Lipids Health Dis 16:151. https://doi.org/10.1186/s12944-017-0539-x
Rajathi AA, Sundarraj AA, Leslie S et al (2017) Processing and medicinal uses of cardamom and ginger—a review. J Pharm Sci Res 9(11):2117–2122
Rani B, Saraswat R, Prasad M et al (2018) Cardamom (*Elettaria cardamomum*): a spice of prominent healthcare. J Biol Chem Res 35:194–199
Ravindran PN (2002) Introduction. In: Ravindran PN, Madhusoodhanan KJ (eds) Cardamom—the genus *Elettaria*. Taylor and Francis, pp 1–10
Robert JL (ed) (1986) Le livre des épices et herbes. Flammarion, Paris
Rodríguez P, Sierra W, Rodríguez S et al (2006) Biotransformation of 1,8-cineole, the main product of Eucalyptus sp. oils. Electron J Biotech 9(3):232–236
Rosengarten F Jr (1969) The book of spices. Jove Publ., Inc., New York, pp 23–96
Sahadevan PC (1965) Cardamom. Farm bulletin no. 4. Department of Agriculture, Kerala, India, p 90
Sangun MK, Aydin E, Timur M et al (2007) Comparison of chemical composition of the essential oil of *Laurus nobilis* L. leaves and fruits from different regions of Hatay, Turkey. J Environ Biol 28(4):731–733
Santos FA, Rao VS (2001) 1,8-cineole, a food flavoring agent, prevents ethanol-induced gastric injury in rats. Dig Dis Sci 46(2):331–337
Savan EK, Kucukbay FZ (2013) Essential oil composition of *Elettaria cardamomum* Maton. J Appl Biol Sci 7(3):42–45
Sengupta A, Bhattachanjee S, Aggarwal B et al (2009) Cardamom (*Elettaria cardamomum*) and its active constituent, 1,8-cineole. World Scientific Publishing, Singapore

Sharma S, Sharma J, Kaur G (2011) Therapeutic uses of *Elettaria cardamomum*. Int J Drug Formul Res 2:102–108

Sharma B, Vasudeva N, Sharma S (2020) Chemical composition and anti-scabies activity of essential oil of *Elettaria cardamomum* Maton leaves. Acta Pharm Sci 58:192–203

Singh VB, Singh K (1996) Spices. New Age International Ltd, New Delhi, p 253

Singh G, Kiran S, Marimuthu P et al (2008) Antioxidant and antimicrobial activities of essential oil and various oleoresins of *Elettaria cardamomum* (seeds and pods). J Sci Food Agric 88:280–289

Singh HP, Parthasarathy VA, Anandaraj M (2012) Introduction to Zingiberaceae crops. In: Singh HP, Parthasarathy VA, Kandiannan K, Krishnamurthy KS (eds) Zingiberaceae crops—present & future. Westville Publishing House, pp 1–18

Sontakke MD, Syed HM, Sawate AR (2018) Studies on extraction of essential oils from spices (cardamom and cinnamon). Int J Chem Stud 6:2787–2789

Subaddarage J, Sarath K, Vellupillai PE et al (1985) Some studies on the effect of maturity and storage on the chlorophyll content and essential oils of the cardamom fruit (*Elettaria cardamomum*). J Sci Food Agric 36:491–498

Sudharshan MR, Kuruvilla KM, Madhusoodanan KJ (1991) A key to the identification of types in cardamom. J Plant Crops 18:52–55

Thomas E, Jaleel K, Zachariah TJ et al (2006) Comparative quality characterization and molecular profiling of Indian, Sri Lankan and Guatemalan cardamoms. J Food Agric Environ 4(2):129–133

Watt G (1872) Dictionary of economic products of India, vol 3, pp 227–236

Winarsi H, Sasongko ND, Purwanto A et al (2014) Effect of cardamom leaves extract as antidiabetic, weight lost and hypocholesterolemic to alloxan-induced Sprague Dawley diabetic rats. Int Food Res J 21:2253–2261

Zachariah TJ (2002) Chemistry of cardamom. In: Ravindran PN, Madhusoodanan KJ (eds) Cardamom—the genus *Elettaria*. Taylor and Francis, London, pp 69–90

Genetic Resources of Cardamom: A Global Perspective

2

P. M. Priyadarshan and K. K. Sabu

Abstract

Cardamom, one of the oldest spices globally, believed to be originated from India and Sri Lanka. Guatemala now stands as the largest producer of commercial cardamom. The reproductive biology of cardamom reveals its herbaceous perennial nature, propagated through seeds or rhizomes. Cultivation and breeding efforts have led to the development of distinct varieties like Malabar, Mysore, and Vazhukka, each suited to different altitudes and environmental conditions. Genetic diversity plays a crucial role in cardamom's agronomic properties, with ongoing breeding programs focused on enhancing yield, disease resistance, and other desirable traits. Looking ahead, climate change poses challenges to cardamom cultivation, emphasizing the need for research into its genetic adaptation to varying altitudes and environmental conditions. Genomic studies offer promise for developing high-yielding, stress-tolerant varieties/clones, ensuring the sustainability of cardamom farming and its continued contribution to global spice production and economic prosperity. In the days to come, utilization of Crop Wild Relatives (CWR) is vital to introducing new genetic variation to any cop species, and cardamom is exception.

Keywords

Cardamom production · Cultivation · Genetic diversity · Reproductive biology · Climate change

2.1 Introduction

Cardamom is thought to be one of the oldest spices in the world. This spice has been used for at least 4000 years. Cardamom was employed by the ancient Egyptians for a variety of medical uses, ceremonies, and even embalming. Because of its strong scent, cardamom was employed by the Greeks and Romans. During their travels, Vikings originally found this spice and returned it to Scandinavia. Numerous literature on Indian mythology refer to cardamom as a spice and remedy. Cardamom is mentioned as an ingredient in several remedies in the Charaka Samhitā, an Indian medicinal compendium

P. M. Priyadarshan (✉)
(formerly with the Rubber Research Institute of India), 9, Neelanjali, Kairali Nagar, ChirakkaraTemple Road, Kaimanam, Thiruvananthapuram 695018, India
e-mail: rriipriya@gmail.com

K. K. Sabu
KSCSTE—Jawaharlal Nehru Tropical Botanic Garden and Research Institute, Palode, Thiruvananthapuram, Kerala 695562, India

K. K. Sabu and C. Kole (eds.), *The Cardamom Genome*, Compendium of Plant Genomes,
https://doi.org/10.1007/978-3-031-71829-8_2

published during the second and third centuries AD. Additionally, cardamom is mentioned as an offering during ceremonies in Sanskrit manuscripts from the fourth century BC, specifically in Kautilya's Arthashasthra.

Cardamom was added to the *Manasollasa*, or Book of Splendour, a list of components for *panchasugandha-thambula*, or five-fragrance betel chew, in India during the eleventh century. It was used in recipes at the Sultan of Mandu's court starting around 1500. Since Arab traders introduced true cardamom into the mainstream a millennium ago, it has been a traded commodity with South Asia. Portuguese traveler Barbosa wrote in 1524 of exports from the Malabar Coast of wild cardamom being grown there. The international cardamom trade was well established by the time Garcia de Orta, a Portuguese Jewish physician who worked primarily in Goa and was a pioneer in the fields of tropical medicine, pharmacognosy, and ethnobotany, arrived in 1563. Cardamom capsules have been used for centuries in traditional medicine and cooking. These uses include treating digestive and kidney disorders, asthma, gum and tooth infections, cardiovascular, pulmonary, kidney, and lung-related disorders (Vaidya and Rathod 2014).

The original source of cardamom (*Elettaria cardamomum* (L.) Maton; Zingiberaceae) was wild plants found in Southern India's Western Ghats. Because cardamom was so abundant in this area, it was given the name Cardamom Hills. According to Purseglove et al. (1981), cardamom is thought to be native to Sri Lanka and South India. British colonists established cardamom plantations in the nineteenth century. The largest producer of cardamom for commercial use is Guatemala. According to Adegoke et al. (1998) and Garg et al. (2016), cardamom is grown in Guatemala, India, Sri Lanka, Tanzania, El Salvador, Vietnam, Laos, Indonesia, Costa Rica, Mexico, Cambodia, Nepal, and Papua New Guinea.

Cardamom is grown in three southern Indian states (Kerala, Karnataka, and Tamil Nadu) at elevations between 900 and 1400 m MSL. It is mostly grown in the Indian Cardamom Hills of Kerala, which are 1050 km^2 in size and are recognized as Cardamom Hill Reserves (Madhusoodanan et al. 2002) (Fig. 2.1). According to reports (Mayne 1951; Abraham and Tulasidas 1958; Govindarajan et al. 1982), cardamom exhibits a high degree of genetic diversity in India. Due to its cross-pollination, cardamom exhibits a wide range of phenotypic variations in the natural world. Among the more noticeable variations are compound panicles, branched cincinnus, sterility in females, and cleistogamy (Prasath and Venugopal 2004; Sudharshan and Bhat 1998; Madhusoodanan et al. 1994).

Oscar Majus Kloeffer brought cardamom to Guatemala for the first time in 1914. Elevations

Fig. 2.1 View of cardamom plantation with shade trees

between 800 and 1300 m MSL are suggested by Guatemala as being optimal for manufacturing. Cardamom prefers 1000–3500 mm of rainfall per year, and it grows well in regions of Guatemala that receive up to 5000 mm of rain annually (Bentley 2014; USDA-FAS 2014)). Cultivating cardamom requires evenly spaced rainfall. Guatemalan growing regions receive 170–200 days of rain annually. Cardamom also needs careful management of 40–60% shade because it is intolerant of intense sunlight. Cardamom is grown in Guatemala in cooler climates at higher altitudes and with less shadow. Compared to India, Guatemala has a far more favorable rainfall pattern. There are two peaks to the yearly distribution of rainfall (2000–5000 mm) (Korikanthimath 2002). The lack of heat and drought stress has resulted in significantly higher cardamom yields in Guatemala, with a national average of 300 kg/ha. A similar scenario with larger yields also exists in Papua New Guinea (Krishna 1997).

The average acreage, production, and productivity of cardamom in India over a 35-year period were found to be 100.048 thousand ha, 12,982.35 metric tons, and 0.128 MT/ha, respectively (Bagalkoti et al. 2019). According to Spices Board, India's statistics, productivity for the 2020–21 fiscal year was 22,520 T kg (see https://www.indianspices.com). According to Statista (https://www.statista.com), in 2017–18, the production was 33,000 MT. In any case, this chapter is not intended to provide a comprehensive overview of cardamom yield, production, and agronomy.

2.1.1 Reproductive Biology

A herbaceous perennial, cardamom can reach heights of two to five meters. Seeds or the vegetative division of underground rhizomes are the two methods of propagation. Encircling the leaf sheaths forms the aerial stem. The lanceolate leaves are 30–35 cm long and 7–10 cm wide, with an acuminate apex and a dark green color. From the axils of subterranean stems, tillers protrude (Fig. 2.2). Monsoon seasons are

Fig. 2.2 Typical cardamom plant

when the majority of vegetative buds are formed (Pruthi 1992). The panicles known as inflorescences emerge from the rhizomes and have a long, cane-like peduncle with internodes and nodes. Typically, the enlarged bases of tillers produce two to four panicles, while certain genotypes produce panicles with several branches. The majority of cardamom varieties have white flowers with pink lines streaking the center lip (Sudharshan et al. 1991; Telja et al. 2006).

There is a scale leaf at each node of the inflorescence, and flowers are carried on a modified helicoids cyme (cincinnus, according to Holttum 1950). As a result, panicles are branching (Madhusoodanan et al. 2002). Certain compound panicle varieties have a direct impact on yield. These compound panicles appear in populations of segregant seedlings. In order to use agricultural genetic resources for crop development programs, it is necessary to pay attention to the natural segregants (Pattanshetty and Prasad 1976; Parameswar 1973). On panicles generated both that year and the year before, blossoming typically occurs throughout the

Fig. 2.3 **a** Flower of cardamom. **b** Honey bee pollinating flower

entire year. India's cardamom flowering season spans six months, from May to October, during which most cardamom-growing regions experience the southwest monsoon. June through August sees the production of about 75% of blooms (Madhusoodanan et al. 2002).

The blossoms exhibit bisexuality. The oval labellum has three separate lobed areas. According to Priyadarshan et al. (1992), the corolla has three lobed unequally, with the largest lobed portion on the posterior side (Fig. 2.3a). The calyx is tubular. Fertile stamen is extended into a brief crest but joined without connective appendages. The stigma is either below or above the crest. The two-lobed anthers dehisce vertically and adhere to the filament. The pollen grains range in diameter from 75 to 120 μm. The stigma has cilia encircling a tiny cavity in the shape of a funnel. With axial placentation, the ovary is inferior, trilocular, and contains several ovules in each carpel. The largest pollen bursting happens between 5.30 and 6.30 a.m., with anthesis normally commencing at 3.30 a.m. and lasting until 7.30 a.m. (Parameswar 1973; Parameswar and Venugopal 1974; Sinu and Shivanna 2007). The majority of flower pollination occurs by cross-pollination, and over 90% of flower pollination is achieved by honey bees (*Apis cerana*, *A. indica*, and *A. dorsata*) (Fig. 2.3b).

There will be about 60,000 flowers accessible for pollination each day if a cardamom population of 3000 plants/ha is planted at a distance of 1.8 m. Assuming that each bee colony will have roughly 5000 foragers, three colonies/ha will suffice to pollinate these flowers (Belavadi and Parvathi 1993; Belavadi and Parvathi 1998). For seed production gardens, an isolation distance of 15 m is recommended based on pollinator activity (Parvathi et al. 1993). When dried, the fruits (capsules) are meaty and leathery, ellipsoidal or nearly spherical, and non-dehiscent. The fruit is 1–2 cm long and green in color, turning golden yellow when ripe (Fig. 2.4). Up to August, there is an increase in capsule development; after that, it steadily decreases. From flower bud initiation, it takes 25–35 days to reach full bloom stage. After reaching full bloom, capsules develop in approximately 120 days (Krishnamurthy et al. 1989). The commercial commodity is the lot of dried capsules that have been cured in the smoke house (Fig. 2.5). When seeds are young, they are white; as they mature, they become brown and finally black. Each capsule holds 12–32 seeds, depending on the genotype; the ripe seeds are covered in a white mucilaginous coat and are black in color (Murugan et al. 2016). Each seed has a thin layer of mucilaginous membrane (aril). One determinant of improved seed set is the existence of honey bee hives. Plant aging is a controlling factor; once a plant reaches its economic life span, its ability to bear fruit will diminish.

According to the type of panicle, cardamom is divided into three categories: Vazhukka (semi-erect panicle), Mysore (erect panicle), and

Fig. 2.4 Cardamom capsules

Fig. 2.5 Dried capsules

Malabar (prostrate panicle) (Ashokkumar et al. 2020). This is a basic description of the three varieties of cardamom:

Malabar: Plants grow up to 2–3 m tall and are best suited at lower elevations between 600 and 1000 m MSL. Panicles are perpendicular to the soil's surface and completely prostrate. Malabar is comparatively less prone to shoot borer and thrips. In Tamil Nadu and Kerala, it can blossom even with little rainfall (Fig. 2.6). The majority of cured capsules are spherical and measure about 18 mm in length.

Mysore: This type can withstand elevations between 900 and 1200 m MSL. When fully grown, the sturdy plants reach a height of 3–5 m (Fig. 2.7). Panicles are perfectly upright and

Fig. 2.6 Malabar type

Fig. 2.7 Mysore type

most suited to areas like Kerala and Karnataka with evenly spaced rains. The ribbed, three-cornered cured capsules are 21 mm in length and are typically a little bit longer than Malabar capsules.

Vazhukka: This type possesses intermediate traits from both the Malabar and Mysore varieties, resulting from a spontaneous hybridization between both. It is ideally suited to altitudes between 900 and 1200 m above mean sea level. The panicles are semi-erect, and the plants are sturdy (Fig. 2.8). Capsules have an ovoid or globose form. These kinds' distinguishing characteristics are listed in Table 2.1. Nair (2006) lists three additional varieties of cardamom:

Mysorensis: A sturdy, towering plant with pubescent or glabrous leaves. There are flexous panicles on this kind. Short racemes of flowers are formed. The capsules are clearly three-angled and bold.

Laxiflora: Not as sturdy as the Mysorensis type, but not as tall either. The petioles of glabrous leaves are short. The decumbent panicles

Fig. 2.8 Vazhukka type

of this kind are flexuous and loose. In four to forty brief, loose racemes, the blooms are produced. The fusiform, oblong-oblong capsules vary in shape. There are several other known cardamom cultivars in India. They can generally be categorized as ecotypes of var. Vazhukka, var. Mysore, or var. Malabar. Bijapur, Kannielam, Makaraelam, Munjarabad, Nadan, and Thara are the most prevalent among them.

The wild cardamom from Sri Lanka (*E. ensal* Abheywickrama). There is considerable disagreement regarding the botanical identity of the Indian types mentioned above as well as the wild cardamom found in Sri Lanka. Different authors have given different names to different types of cardamom, including: *E. cardamomum* var. *minus*, *E. cardamomum* var. *miniscula*, *E. cardamomum* var. *major*, *E. cardamomum* var. *majus*, and *E. cardamomum* var. *minor*.

Several more varieties, such as *mysorensis* and *laxiflora*, have been identified by morphological characteristics. Madhusoodanan et al. (2002) also mention Sri Lankan wild cardamom (*E. ensal* Abheywickrama). Even as early as 1912, Ridley described the two types of Sri Lankan cardamom: var. *minus* and *majus*. But currently, Malabar, Mysore, and Vazhukka are the most common varieties worldwide.

Elettaria cardamomum belongs to a single species group. Its balanced tetraploid nature is indicated by the basic chromosomal numbers, $x = 12$ and $2n = 48$ (Madhusoodanan et al. 2002). Only *E. cardamomum* is significant economically, according to Holttum (1950), Mabberley (1987), and Wills (1967). The seven species of the genus *Elettaria* are spread throughout India, Sri Lanka, Malaysia, and Indonesia. Nair (2006) contends, however, that one of the largest species of the genus grown in Malaysia is *E. longituba* Holttum (Syn. *E. longituba*) (Fig. 2.9). Its flowers are solitary

Table 2.1 Difference between types of cardamom

Parameters	Malabar	Mysore	Vazhukka
Adaptability	Lower elevation (600–1000 m MSL)	Higher elevation (900–1200 m MSL)	Higher elevation (900–1200 m MSL)
Tolerance to drought	With stands long dry spell (4–6 months)	Needs well distributed rain	Needs well distributed rain
Plant stature	Dwarf (2–3 m)	Tall (3–5 m)	Tall (3–5 m)
Leaf	Short petiole	Long petiole	Long petiole
Panicle	Prostrate	Erect	Semi-erect
Bearing nature	Early, short span of flowering	Late, long flowering span	Late, long flowering span
Capsule color	Pale/golden yellow	Green	Green

Source Sudharshan et al. (1991)

Fig. 2.9 *Elettaria longituba*

and occur at longer intervals; each cincinnus has a small number of blossoms. The cincinnus seems to quit flowering as soon as fruit starts to emerge. Despite their size, the fruits are not valuable commercially (Holttum 1950). Six species of *Elettaria* were described by Sakai and Nagamasu (2000) in their research on Bornean Zingiberaceae: *E. kapitensis*, *E. surculosa*, *E. linearicrista*, *E. longipilosa*, and *E. brachycalyx*. There are thirteen species of *Elettaria*, according to The Plant inventory (TPL), a working inventory of all known plant species created by the botanical community in response to Target 1 of the Global Strategy for Plant Conservation (GSPC) 2002–2010. Although TPL has not changed since 2013, it served as the foundation for the World Flora Online's Taxonomic Backbone (WFO). The most recent data is available at www.worldfloraonline.org (Table 2.2).

Within the Zingiberaceae family are a few related genera. Take *Amomum*, *Aframomum*, and *Alpinia* as examples. Black cardamom, sometimes known as large cardamom, is indigenous to Bangladesh, Sikkim, Nepal, and other Southeast Asian nations (Fig. 2.10). *Aframomum danielli* (Hook. f.) K. Schum., the botanical name for African cardamom, is indigenous to southeast Africa, particularly Tanzania, Cameroon, Madagascar, and Guinea (Fig. 2.11). *Alpinia* sp. are native of PNG and East Indies (Fig. 2.12). However, due to exorbitant pricing, only Asian nations are currently involved in international commerce of both tiny and giant cardamom. After saffron and vanilla, cardamom

Table 2.2 Species enumerated under Plant List[a]

Name	Status	Confidence level	Source
Elettaria alba Blume	Unresolved	⁎⁎⁎	WCSP
Elettaria brachycalyx S. Sakai and Nagam.	Accepted	⁎⁎⁎	WCSP
Elettaria brasiliensis Raddi	Unresolved	⁎⁎⁎	WCSP
Elettaria cardamomum (L.) Maton	Accepted	⁎⁎⁎	WCSP
Elettaria ensal (Gaertn.) Abeyw.	Accepted	⁎⁎⁎	WCSP
Elettaria kapitensis S. Sakai and Nagam.	Accepted	⁎⁎⁎	WCSP
Elettaria linearicrista S. Sakai and Nagam.	Accepted	⁎⁎⁎	WCSP
Elettaria longipilosa S. Sakai and Nagam.	Accepted	⁎⁎⁎	WCSP
Elettaria longituba (Ridl.) Holttum	Accepted	⁎⁎⁎	WCSP
Elettaria multiflora (Ridl.) R. M. Sm.	Accepted	⁎⁎⁎	WCSP
Elettaria rubida R. M. Sm.	Accepted	⁎⁎⁎	WCSP
Elettaria stoloniflora (K. Schum.) S. Sakai and Nagam.	Accepted	⁎⁎⁎	WCSP
Elettaria surculosa (K. Schum.) B. L. Burtt and R. M. Sm.	Accepted	⁎⁎⁎	WCSP

[a] ***The Plant List*** (TPL) was a working list of all known plant species produced by the botanical community in response to Target 1 of the 2002–2010 Global Strategy for Plant Conservation (GSPC). TPL has been static since 2013, but was used as the starting point for the Taxonomic Backbone of the ***World Flora Online (WFO)***, and updated information can be found at www.worldfloraonline.org. *WCSP* World Checklist of Selected Plant Families

Fig. 2.10 Large cardamom (*Amomum subulatum*)

Fig. 2.11 *Aframomum danielli* (Hook. f.) K. Schum.

is the third most costly spice in the world and is known as the "queen of spices" due to its pleasing perfume and flavor. PNG and the East Indies are Alpinia's native lands.

2.2 Phenotypic and Genotypic Diversity

Rich genetic variety is responsible for the diverse agronomic, production, and quality properties of cardamom. IPGRI issued a comprehensive description on cardamom germplasm (IPGRI 1994). There is significant variation in panicle length and yield/plant, according to Prasath et al. (2001). Significant differences exist in characteristics such as inflorescence branching, fruit (capsule) size, form, pubescence of leaves and plants, and preservation of green color (Madhusoodanan et al. 1994). The volatile oil in cardamom, which ranges from 6.5 to 10.5%, is used. According to Zakir (2019), several accessions (Acc. 221 and Acc. 218) have an essential oil content of 7.8%. Oil has a low concentration of 1,8-cineole and a significant component of aroma-bearing compounds such as linalyl and alpha terpinyl acetates (Zachariah et al. 1998). Out of 134 disease escapes gathered

Fig. 2.12 Allied genera **a** *Aframomum melegueta*; **b** *Aframomum angustifolium*; **c** *Amomum villosum*; **d** *Aframomum zambesiacum*; **e** *Alpinia purpurata*; **f** *Alpinia speciosa*

from hot locations, 17 accessions resistant to the mosaic (katte virus) illness were found (Venugopal 1999). The cardamom capsules from Guatemala are much larger (Fig. 2.13).

The goal of cardamom breeding is to produce plants with greater yields but also with resistance to biotic stresses like "katte" and fungal diseases like rhizome rot, clump rot, and capsule rot, as well as drought tolerance. Preferable characteristics include bold capsules with more seeds/fruit, a larger percentage of capsule dry recovery (> 22%), a higher percentage of essential oils, α-terpenyl acetate, which gives the capsule its flavor and scent, and varieties with a wide range of adaptation (Zakir 2019). Primary selections from germplasm, from open-pollinated progenies and hybrids, are followed in cardamom breeding (Nair 2020). A total of eighteen cardamom cultivars with high yields were made available for cultivation (Table 2.3). While IISR Avinash and ICRI 4 are somewhat resistant to rhizome rot, IISR Vijetha is a line

Fig. 2.13 a Indian, b Guatemalan

that is resistant to the katte virus. The capsules in PV 1 are bold and lengthy. There are types with multiple-branched panicle (Fig. 2.14) and pink tillers (Fig. 2.15).

A long, cane-like peduncle with nodes and internodes is characteristic of a typical inflorescence or panicle (Fig. 2.16). Every node has a scale leaf, and the modified helicoids cyme (cincinnus) bears flowers in the axil. As a result, Madhusoodanan et al. (2003) describe the panicle as branching. Certain cultivars' seedling progenies have many branching panicles (Fig. 2.14) (Prasath and Venugopal 2009). Here, further branching (secondary and tertiary branches) occurs on the central peduncle, resulting in several branching panicles (compound panicles). In addition to synchronized flowering and fruit set, multiple-branched panicle varieties have a high yield potential per unit area when compared to conventional panicle kinds (Prasath and Venugopal 2009).

The cardamom gene pool is being eroded by changes in the Western Ghats of south India, which necessitates methodical germplasm collection and research. A sizable gene pool is kept up to date at the Indian Cardamom Research Institute in Myladumpara, the Indian Institute of Spices' regional research station in Appangala, the Cardamom Research Station in Pampadumpara (Kerala Agricultural University), and the centers of the Indian Council of Agricultural Research's All India Coordinated Research Project on Spices. Clonal repositories, a crop cultivated vegetatively, safeguard germplasm. This requires a lot of labor and is vulnerable to risks including pest outbreaks, droughts, and illnesses like "katte" and rhizome rot, which lower productivity. Thus, germplasm collections would be safe if preserved in vitro (Nirmal Babu et al. 1999). It was shown that vitrified shoots and encapsulated embryos may be stored for an extended period of time using cryopreservation (liquid nitrogen at a temperature of 1960 °C) (Ravindran et al. 2004). But before being applied widely, these protocols need to be refined even more.

Although there is noticeable phenotypic diversity between the many cardamom varieties, this variation does not show up clearly when data are analyzed using the right statistical tools. This suggests that the genotypes of the three cultivar groups (Mysore, Malabar, and Vazhukka) may have a shared ancestral and close relationship and that genetic divergence in cardamom is not solely caused by geographic origin (Prasath and Venugopal 2004). A cluster study using 310 accessions of different cardamom varieties gathered from diverse cardamom pockets in south India showed that clusters 1 and 2 had the greatest inter-cluster distance (4.32), followed by clusters 2 and 3 (3.463). There was a significant level of genetic variation among genotypes, as evidenced by the inter-cluster distances being larger than the intra-cluster distances. The same cultivar group's accessions were dispersed over various clusters. For Malabar, Mysore, and Vazhukka, there is no clear grouping of accessions (Prasath and Venugopal 2004).

By selecting for desired traits like greater yield and better capsule characteristics, all of the current enhanced cultivars have evolved. In order to verify the superiority of the chosen clone, cardamom selection is based on both qualitative and quantitative traits from

Table 2.3 Released varieties of cardamom

S. No.	Variety	Developed by	Year	Pedigree/parentage	Yield (kg/ha)	Recovery (%)	Oil %	Adaptable to	Salient features
1.	ICRI 1	ICRI (Spices Board), Myladumpara, Kerala	1992	Primary clone from Chakkupalam collection, Malabar type	325	22.90	8.70	South zone of Idukki, Kerala	An early maturing; globose, round and extra bold dark green capsules; medium-sized panicle with profusely flowering, early maturing type
2.	ICRI 2	ICRI (Spices Board), Myladumpara, Kerala	1992	Primary clone Mysore type	375	22.50	6.67	Parts of Kerala and Tamil Nadu	Adaptable to high altitude and irrigated conditions; medium long panicles, oblong bold and parrot green capsules, tolerant to azhukal (rhizome rot) disease
3.	ICRI 3	ICRI (Spices Board), Myladumpara, Kerala	1994	Primary clone Malabar type	440	22.00	6.60	Tracts of Karnataka	Early maturing; long pubescent leaves; tolerant to rhizome rot disease; oblong, bold parrot green capsules; suitable to hill zone of Karnataka
4.	ICRI 4	ICRI (Spices Board), Myladumpara, Kerala	1997	Primary clone from lower pulleys Malabar type	455	22.76	6.40	Lower Palani hills, Tamil Nadu	Early maturing; medium-sized panicles, globose bold capsules; for low rainfall areas, relatively tolerant to rhizome rot and capsule borer
5.	ICRI 5	ICRI (Spices Board), Myladumpara, Kerala	2006	Hybrid of MCC 260 × MCC 49	1543	23.15	7.13	Kerala and parts of Tamil Nadu	First hybrid variety; early maturing; moderately tolerant to drought; high yield under intensive management; capsule size 68%; more than 7 mm

(continued)

Table 2.3 (continued)

S. No.	Variety	Developed by	Year	Pedigree/parentage	Yield (kg/ha)	Recovery (%)	Oil %	Adaptable to	Salient features
6.	ICRI6 (MCC-73)	ICRI (Spices Board), Myladumpara, Kerala	2006	Primary clone of (acc. MCC-73)	1200	19.00	7.33	Kerala and parts of Tamil Nadu	High yield; medium maturity; relatively tolerant to drought; high percentage of bold capsules; capsule size 71%; more than 7 mm
7.	ICRI 7	ICRI (Spices Board), Myladumpara, Kerala	2010	Hybrid	–	22.00	8.84	Wayanad, Kerala	Suitable to Wayanad; Kerala, semi-erect panicles; angular bold capsules; oleoresin 7.99%
8.	ICRI 8	ICRI (Spices Board), Myladumpara, Kerala	2014	Primary clone from Sakleshpur with oval, bold and pale green capsules	655	–	7.10	Tracts of Karnataka	Suitable to Karnataka
9.	IISR Kodagu Suvasini (CCS-1)	Indian Institute of Spices Research, Calicut, Kerala	1997	Selection from open-pollinated progeny of CL-37 from RRS Mudigere Malabar type	745	22	8.70	Kodagu, Hassan and Chikmagalur	Early maturing; suitable to high density planting; long panicle; tolerant to rhizome rot, thrips, shoot/panicle/capsule borer
10.	IISR Avinash (RR-1)	Indian Institute of Spices Research, Calicut, Kerala	2001	A selection from OP progeny of CCS-1, a Malabar type	847	20.8	6.7	For hotspots of rhizome rot prone areas	Dark green capsules; tolerant to rhizome rot, shoot/panicle/capsule borer
11.	IISR Vijetha (NKE-12)	Indian Institute of Spices Research, Calicut, Kerala	2001	Primary clone from field-resistant plants of Katte Malabar type	643	22	7.9	Kodagu, Hassan, Chikmagalur and North Wayanad; for Katte prone areas. For moderate rainfall and moderate to high shade areas	Resistant to *katte* virus; bold capsules; field tolerance to thrips and borer

(continued)

Table 2.3 (continued)

S. No.	Variety	Developed by	Year	Pedigree/parentage	Yield (kg/ha)	Recovery (%)	Oil %	Adaptable to	Salient features
12.	Mudigere 1	Regional Research Station, UAS, Mudigere, Karnataka	1984	Clonal selection from Malabar type	300	20	8	Malanad areas of Karnataka, under rainfed conditions	Erect and compact panicle; suitable to high density planting; moderately tolerant to thrips, hairy caterpillar and white grubs, pubescent lea; short panicle, pale green, oval bold capsule
13.	Mudigere 2	Regional Research Station, UAS, Mudigere, Karnataka	1994/1996	Primary clone from open-pollinated seedlings Malabar type	475	–	8	Malanad areas of Karnataka, under rainfed conditions	Early maturing; suitable to high density planting; round/oval bold capsules
14.	PV 1	Cardamom Research Station, KAU, Pampadumpara, Idukki, Kerala	1991	A selection from Walayar collection Malabar type	260	19.9	6.8	All tracts of Kerala and parts of Tamil Nadu	Early maturing; short panicle, elongated slightly ribbed light green capsules; long, bold capsule
15.	PV 2	Cardamom Research Station, KAU, Pampadumpara, Idukki, Kerala	2001	Selection from open-pollinated seedlings of PV-1 Malabar type	982	23.8	10.45	All tracts of Kerala and parts of Tamil Nadu	Early maturing; lengthy panicle; long bold capsules; high dry recovery percentage; field tolerance to stem borer and thrips; suitable to 1000–1200 m MSL
16.	PV 3	Cardamom Research Station, KAU, Pampadumpara, Idukki, Kerala	2018	Malabar type	416	18.5	–	For partial shade (50–60%) in tropical evergreen forests	Drought tolerant; moderately tolerant to capsule borer; ellipsoid capsule; number of capsules—342/panicle
17.	PV 5	Cardamom Research Station, KAU, Pampadumpara, Idukki, Kerala	2018	Vazhukka type	594	18	–	Tracts of Kerala	Tolerant to thrips; ovoid capsule; number of capsules 153/panicle

(continued)

Table 2.3 (continued)

S. No.	Variety	Developed by	Year	Pedigree/parentage	Yield (kg/ha)	Recovery (%)	Oil %	Adaptable to	Salient features
18.	Njallani	Farmers selection popularly known as Njallani green gold	1990s	Selection from Vazhukka type	1600	–	9.01	All tracts of Kerala and parts of Tamil Nadu	Non-pubescent; semi-erect; globose extra bold and dark green capsules; high yielder

Courtesy: Cardamom cultivation practices Spices Board 2009, AICRPS—http://aicrps.res.in/sites/default/files/inline-files/Varieties.pdf, www.kau.in

preliminary, comparative yield testing and multi-location studies. The goal of inter-varietal hybridization was to create lines with high yield, drought tolerance, and "katte" resistance by crossing selected superior cultivars. These cultivars are undergoing on-farm trials. At the Indian Institute of Spices Research, Appangala, numerous crosses have been created to combine high production with resistance to cardamom mosaic and rhizome rot diseases. These crosses are presently being evaluated. Cardamom hybrid seedlings showed varying degrees of considerable positive heterosis (Padmini et al. 2001). Fifteen hybrid combinations are shortlisted for additional assessment based on their individual performance, heterosis, and combining ability. Non-additive gene activity affected plant height, total tillers, bearing tillers, and yield per plant (Prasath and Venugopal 2001).

2.3 Cardamom of Guatemala

Cardamom grows well in Guatemala at elevations between 250 and 1600 m above sea level and between 10 and 35 °C. Vazhukka is more common than the other two varieties (Mysore and Malabar). Elevations between 760 and 1400 m are advised by the Asociación de Cardamomeros de Guatemala (Cardegua), with 800–1300 m being the optimal range (Anonymous 2014). Despite being produced in regions with 5000 mm of rainfall annually, the ideal rainfall range is 1000–3500 mm (with 170–200 days) (Bentley 2014). A careful 40–60% shadow management strategy is needed. Lower temperatures offset the effects of growing in less shade at higher elevations (Anonymous 2014).

With about 43,000 cardamom-producing farms (fincas), the department of Alta Verapaz in north central Guatemala produces over 70% of the country's total output. In addition, there are noteworthy growing regions in Huehuetenango, Quiché, Baja Verapaz, and the hilly regions of Izabal; 14 additional departments have less than 100 farms with smaller output levels. In agroforestry systems, cardamom has the potential

Fig. 2.14 Multiple-branched panicle (courtesy: ICRI, Myladumpara)

Fig. 2.15 Pink tiller type (courtesy: ICRI, Myladumpara)

to be highly valuable, particularly in areas with marginal or sharply sloping terrain. The majority of the cardamom plant's shallow roots reach a depth of about 40 cm. With the majority of the lateral roots located within a radius of 50–75 cm, the root system is likewise somewhat narrow. They do not compete with shade trees for nutrients because of their shallow roots. Compared to Indian cardamom, the genetic diversity of Guatemalan cardamom is low grade and poorly understood (Bentley 2014).

2.4 Prospectives

India, along with the depth of its cultural heritage, is also praised worldwide for the aroma of its spices. Cardamom, native to India and Sri Lanka, is believed to have its root origin in the tropical evergreen woods of the Western Ghats in South India and is extensively in cultivation globally. Cardamom has supported in enhancement of the financial status and standard of living of farmers in high-altitude regions. Climate change is currently having a severe impact on crop productivity, destroying the intermolecular connections required for healthy growth and leading to large yield losses with serious implications for future international food security. It is found that at varying elevations, distinct environmental conditions seem to manipulate the physiological performance of cardamom, which regulates its yield and other significant agronomic traits. A comparative study of genomics at diverse altitudes may provide knowledge

Fig. 2.16 A typical cardamom panicle

of the molecular genetic mechanism underlying the adaptation of cardamom to high elevations. Developing novel high-yielding cardamom varieties tolerant to abiotic stress could provide a substantial livelihood to the farmers in addition to its contribution to the gross production of cardamom. Plant genomics research has extensively brought about a revolution in the field of crop improvement. A profound understanding of the genetic and molecular characteristics of a crop plant from its wild and cultivar lineage will illustrate a path to either upgrade an established genotype or to develop a novel variety suitable for cultivation in different climatic conditions, preferably be able to resist abiotic stresses. With the use of contemporary breeding techniques, it is possible to enrich crop genetic resources by introducing adaptive traits prevailing in wild relatives and traditional varieties regularly.

Acknowledgements Thanks are due to Dr. C. Nunes Zuffo, *Center d'etudes et coopération Internationale, Québec, Canada,* Dr. Jeffery Bentley, Agricultural Anthropologist, Agro-Insight, Cochabamba, Bolivia, and Ms. Claudia Calderon, Inter-American Institute for Cooperation in Agriculture (IICA), Guatemala, for providing details of Guatemalan cardamom.

References

Abraham P, Thulasidas G (1958) South Indian cardamom and their agricultural value. In: Technical Bulletin, vol 79. ICAR, New Delhi, pp 1–27

Adegoke GO, Vijaykumar M, Gopalakrishna AG, Varadaraj MC, Sambaiah K, Lokesh BR (1998) Antioxidants and lipid oxidation in foods-a critical appraisal. J Food Sci Technol 35(4):283–298

Anonymous (2014) Rapid economic feasibility in Guatemala: cardamom, cinnamon, macadamia nut, nutmeg. Catholic Relief Services, Baltimore, USA, 44 pp

Ashokkumar K et al (2020) Botany, traditional uses, phytochemistry and biological activities of cardamom [*Elettaria cardamomum* (L.) Maton]—a critical review. J Ethnopharmacol 246. https://doi.org/10.1016/j.jep.2019.112244

Bagalkoti JB et al (2019) Instability analysis of productivity and production of cardamom. Int J Curr Microbiol Appl Sci 8:1500–1503

Belavadi VV, Parvathi C (1993) Landing and departure routes used by honeybees on cardamom (Elettaria cardamomum Maton) In: Veeresh GK, Uma Shankar R, Ganeshaiah KN (eds) Proceeings of the international symposium on pollination in tropics. IUSSI, Indian Chapter, Bangalore, pp 73–76

Belavadi VV, Parvathi C (1998) Estimation of honeybee colonies required for effective pollination in cardamom. In: Proceedings of the national symposium diversity of social insects & other arthropods and the functioning ecosystems, 30. Mudigere: III Congress of IUSSI, p 30

Bentley JW (2014) Cardamom—a crop cannot run away from its problems. Outlooks Pest Manag. https://doi.org/10.1564/v27_feb_02

Garg G, Sharma S, Dua A, Mahajan R (2016) Antibacterial potential of polyphenol rich methanol extract of Cardamom (*Amomum subulatum*). J Innov Biol 3(1):271–275

Govindarajan VS, Narasimhan S, Raghuveer KG, Lewis YS, Stahl WH (1982) Cardamom—production, technology, chemistry, and quality. Crit Rev Food Sci Nutr 16(3):229–326

Holttum RE (1950) The Zingiberaceae of the MalayPeninsula. Gardens' Bull Singapore 13:1–250

IPGRI (1994) Descriptor for cardamom (*Elettaria cardamomum* Maton). International Plant Genetic Resources Institute, Rome, pp 27–41

Korikanthimath VS (2002) Agronomy and management of cardamom. In: Ravindran PN, Madhusoodanan KJ (eds) Cardamom: the genus Elettaria. CRC Press, pp 107–144

Korikanthimath VS, Mulge R, Hegde R, Hosmani MM (2002) Evaluation of elite lines of cardamom (Elettaria cardamomum) for yield and yield parameters. Indian J Hortic 59(3):326–330

Krishna KVS (1997) Cardamom plantations in Papua New Guinea. Spice India 10:23–24

Krishnamurthy R, Bhagwat K (1989) Polyamines as modulators of salt tolerance in rice cultivars. Plant Physiol 91:500–504

Mabberley DJ (1987) The plant book, a portable dictionary of the higher plants. Cambridge University Press, New York

Madhusoodanan KJ, Kuruvilla, Priyadarshan PM (1994) Genetic resources of cardamom. In: Chadha KL, Rethinam P (eds) Advances in horticulture, vol 9. Plantation and spice crops (part I). Malhotra Publishing House, New Delhi

Madhusoodanan KJ, Pradeep Kumar K, Ravindran PN (2002) Botany, crop improvement and biotechnology of cardamom. In: Ravindran PN, Madhusoodanan KJ (eds) Cardamom—the genus *Elettaria* (medicinal and aromatic plants industrial series). CRC Press, Taylor and Francis Group, London, United Kingdom, pp 11–68

Madhusoodanan KJ et al (2003) Botany, crop improvement and biotechnology of cardamom. In: Ravindran PN, Madhusoodanan KJ (eds) Cardamom—the genus *Elettaria*. Taylor & Francis, London, pp 11–40

Mayne WW (1951) Report on cardamom cultivation in South India. Indian Council of Agricultural Research Bulletin 50:1-80

Murugan M, Dhanya MK, Deepthy KB, Preethy TT, Aswathy TS, Sathyan T, Manoj VS (2016) Compendium on cardamom, 2nd edn. Kerala Agricultural University, Cardamom Research Station, Pampadumpara, Kerala, p 66

Nair KP (2006) The agronomy and economy of cardamom (*Elettaria cardamomum* M.): the "queen of spices." Adv Agron 91:179–471

Nair KP (2020) Cardamom botany. In: Nair KP (ed) The geography of cardamom (*Elettaria cardamomum* M.). Springer, Cham, pp 9–38. https://doi.org/10.1007/978-3-030-54474-4_2

Nirmal Babu K et al (1999) *In vitro* conservation of cardamom (*Elettaria cardamomum* Maton) germplasm. Plant Genet Resour Newslett 11:41–45

Padmini K, Venugopal MN, Ankegowda SJ (2001) Heterosis for seedling characters in cardamom (*Elettaria cardamomum* Maton.). J Spices Aromat Crops 9(2): 145–151

Parameswar NS (1973) Floral biology of cardamom (*Elettaria cardamomum Maton*). Mysore J Agric Sci 7(2):205–213

Parameswar NS, Venugopal R (1974) Study of flowering and anthesis in cardamom (*Elettaria cardamomum Maton*). Mysore J Agric Sci 6:356–361

Parvathi C, Shadakshari YG, Belavadi VV, Chandrappa HM (1993) Foraging behaviour of honey bees on cardamom (Elettaria cardamom Maton). In: Veeresh GK, Umashaanker R, Ganeshaiah KN (eds) Pollination in tropics. IUSSI Indian Chapter, Bangalore, India, pp. 99–103

Pattanshetty HV, Prasad AB (1976) Blossom biology pollination and fruit set in cardamom Elettaria cardamomum. In: Chadha KL (ed) Proceedings of the third international symposium on sub-tropical and tropical horticulture, vol 1. Improvement of horticultural, plantation and medicinal plants Bangalore, India, Feb 8–14, 1972. Today and Tomorrow's Printers and Publishers, New Delhi, India, pp 262–268

Prasath D, Venugopal MN (2001) Genetic analysis for quantitative characters in cardamom (*Elettaria cardamomum* Maton.). Indian J Genet 62:179–180

Prasath D, Venugopal MN (2004) Genetic diversity and conservation of cardamom in India. Plant Genet Resour News Lett 138:55–60

Prasath D, Venugopal MN (2009) Compound inflorescence cardamom (*Elettaria cardamomum* (L.) Maton) in India. Genet Resour Crop Evol 56:749–753

Prasath D, Venugopal MN, Korikanthimath VS (2001) Variability in cardamom (*Elettaria cardamomum* Maton). Indian J Plant Genet Resour 14(2):217–218

Priyadarshan PM, Kuruvilla KM, Madhusodanan KJ, Naidu R (1992) Effects of vivid media protocols and genetic specificity of micropropagation of cardamom (*Eletlaria cardamomum Moton*). In: Subba Rao NS, Balachandram C, Ramakrishnan SV (eds) New trends in biotechnology. Oxford and IBH Pub, pp 109–118

Pruthi JS (1992) Spices and condiments. National Book Trust of India, New Delhi, 287 pp

Purseglove JW, Brown EG, Green CL, Robbins SRJ (1981) Spices, vol 2. Longman, New York, USA

Ravindran PN et al (2004) Conservation of spices genetic resources in *in-vitro* gene banks. ICAR project report. Indian Institute of Spices Research, Calicut, Kerala, India, p 81

Reddy BGS, Siddaramaiah AL, Parameswar NS (1973) Ten minutes pre sowing treatment of cardamom with 50 per cent hydrochloric acid ensures good germination. Curr Res 2:22

Sakai S, Nagamasu H (2000) Systematic studies of Bornean Zingiberaceae: II. Elettaria of Sarawak. Edinb J Bot 57:227–243

Siddagangaiah S, Krishnakumar V, Vadiraj BA, Sudharshan MR (1998) Influence of cultural practices on growth and yield of cardamom (*Elettaria cardamomum Maton*). In: Mathew NM, Jacob CK (ed) Developments in plantation crops research. Proceeding of XII Symposium on Plantation Crops, 1997. Rubber Research Institute of India, Kottayam, pp 195–198

Sinu PA, Shivanna KR (2007) Pollination ecology of cardamom (*Elettaria cardamomum*) in the Western Ghats, India. J Trop Ecol 23:493–496

Sudharshan MR, Bhat SS (1998) Tissue cultured cardamom clones: a comparative study. In: Mathew NM, Jacob CK (ed) Developments in plantation crops research. Proceedings of XII Symposium on Plantation Crops, 1997. Rubber Research Institute of India, Kottayam, pp 73–76

Sudharshan MR, Kuruvilla KM, Madhusoodanan KJ (1991) A key to the identification of types in cardamom. J Plant Crops 18:52–55

Telja R et al (2006) Small cardamom—precious for people, harmful for mountain forests. Possibilities for sustainable cultivation in the East Usambaras, Tanzania. Mt Res Dev 26:131–137

USDA-FAS (2014) Cardamom—the 3Gs—green gold of Guatemala. GAIN report number: GT-1404. USDA Foreign Agricultural Service Global Agricultural Information Network. https://apps.fas.usda.gov/newgainapi/api/report/downloadreportbyfilename?filename=Cardamom%20-%20The%203Gs%20--%20Green%20Gold%20of%20Guatemala_Guatemala_Guatemala_7-11-2014.pdf

Vaidya A, Rathod M (2014) An in vitro study of the immunomodulatory effects of *Piper nigrum* (black pepper) and *Elettaria cardamomum* (cardamom) extracts using a murine macrophage cell line. Int J Res Formal Appl Nat Sci 8:18–27

Venugopal MN (1999) Natural disease escapes as source of resistance against cardamom mosaic virus causing *katte* disease of cardamom (*Elettaria cardamomum* Maton). J Spices Aromat Crops 8:145–151

Wills JC (1967) A dictionary of flowering plants and ferns, 7th edn. Cambridge University Press, New York

Zachariah TJ, Ravindra M, Venugopal MN (1998) Quality of cardamom from different accessions. In: Mathew NM, Jacob CK (eds) Developments in plantation crops research. Allied Pub., India, pp 337–340

Zakir M (2019) Review on breeding method and achievements of cardamom (*Elettaria cardamomum* Maton) and future prospects. Int J Res Agric For 6:16–23

Genetics and Breeding of Cardamom

3

D. Prasath, H. J. Akshitha, and M. S. Shivakumar

Abstract

Cardamom (*Elettaria cardamomum* Maton), a monocot member of the Zingiberaceae, is a well-known spice and medicinal herb since ancient times. The Trade Promotion Council of India estimated the size of the global cardamom industry in value terms as US $69.8 billion in 2018 and is expected to reach USA $86.4 billion by 2025, growing at a CAGR of 2.8%. Global market of cardamom is highly concentrated, and 60% of the production is from two major producing countries, viz. Guatemala and India. Cardamom being cross-pollinated crop offers tremendous scope for increasing the productivity by selection of open-pollinated progenies. Also, amenability for both sexual as well as asexual propagation gives good scope for crop improvement. However, development of improved cultivars in cardamom requires many breeding cycles and dozens of years because of semi-perennial nature of the crop. This chapter highlights the achievements in cardamom breeding for yield, quality, and disease resistance.

Keywords

Pollination · Hybridization · Varieties · Breeding · Variability

3.1 Introduction

The green cardamom, *Elettaria cardamomum* Maton, is a member of the Zingiberaceae family and is widely recognized as the "queen of spices." Second only to saffron in importance and value, cardamom is the dried, mature fruit used in commerce. It is indigenous to South India's Western Ghats, an area of evergreen forests. The plant known as cardamom (Pseophyte) is tolerant of shade and may grow to a height of 600–1200 m above mean sea level. It requires an average annual rainfall of 1500–4000 mm and a temperature range of 10–350 °C. Up until 1979–1980, India was the world's top producer of cardamom. Subsequently, Guatemala became the world's leading manufacturer and exporter of cardamom, contributing to about 90% of the world market. In Sri Lanka, cardamom is also farmed to a lesser amount. Despite its prominence in world trade from time immemorial, cardamom research received much attention in the second half of this century.

D. Prasath (✉)
ICAR-Indian Institute of Spices Research,
Kozhikode, Kerala 673012, India
e-mail: dprasath@gmail.com

H. J. Akshitha · M. S. Shivakumar
ICAR-Indian Institute of Spices Research,
Regional Station, Appangala, Karnataka 571201, India

K. K. Sabu and C. Kole (eds.), *The Cardamom Genome*, Compendium of Plant Genomes,
https://doi.org/10.1007/978-3-031-71829-8_3

3.2 Genetics

3.2.1 Cytogenetics

First cytological study in cardamom was carried out by Gregory (1936) and reported $2n=48$ somatic chromosome number. Chromosomes in cardamom are of intermediate size compared *Hedychium elvesi* and *Costus malarotiendis*. Sharma and Bhattacharya (1959), Chakravarti (1948) and Sudharshan (1989) confirmed the finding of Gregory (1936). Chandrasekhar and Sampath Kumar (1986) reported chromosome number variations in cardamom among the Mysore and Malabar types indicated that aneuploidy as well as chromosomal structural aberrations has contributed to the varietal differentiation. Whereas Anjali et al. (2016a) carried out flow cytometric and chromosomal count studies in wild cardamom and cultivated genotypes which included Mysore, Malabar, and *Vazhukka* types and reported $2n=48$ (tetraploid) with no variation in chromosome number and concluded that all the 30 accessions of cardamom studied are having tetraploid nature. Flow cytometric studies on germplasm accessions (wild, varieties, and landraces) showed low variation for 2C-nuclear DNA value and which was found to be 2.84 pg. The estimated diploid genome size of cardamom was 2783 Mbp (Anjali et al. 2016a) which is comparable range of the Zingiberaceae family (Anjali et al. 2016b).

3.2.2 Genetic Variability and Trait Association in Cardamom

Gopal et al. (1992) found that the number of capsules per plant had a high genotypic coefficient of variation (GCV) and phenotypic coefficient of variation (PCV), followed by the number of panicles and yield per plant. Selection could be aided by traits such genetic progress, heritability, yield/plant, and number of capsules/plant with high GCV (Prasath et al. 2001). High CV for yield per plant and capsule per plant was reported by Prasath et al. (2009a, b). The yield per plant (77.94%) had the highest coefficient of variation, followed by the branches per panicle (49.40%), whereas the dry recovery percentage (7.73%) had the lowest coefficient of variation (Prasath and Venugopal 2009). Senthil Kumar et al. (2017) studied 11 genotypes and observed high CV for number of panicles followed by number of bearing tillers. Hrideek et al. (2015) evaluated 11 genotypes for 32 characters; among the yield characters highest GCV and genetic advance was observed for internodal length, PCV for panicles per clump, and heritability for racemes per panicle. Akhila et al. (2017) observed highest coefficient of variation for wet weight of capsules (63.26%) followed by number of capsules (55.52%).

Nadukeri et al. (2020b) carried out variability studies in 67 cardamom genotypes comprised of germplasm accessions and released varieties and observed highest GCV (23.5%) and PCV (24.3%) for the trait number of leaves per tiller. High heritability (99.0%) and genetic advance over mean (48.2%) for the trait number of vegetative buds per clump was observed. Anisha et al. (2020) studied variability among 12 released varieties and a farmer's popular selection (Njallani) for 14 qualitative traits and observed highest coefficient of variation for number of panicles per plant followed by number of capsules per panicle and lowest CV for dry recovery of capsules followed by panicle per tiller. Sharon et al. (2020) while characterizing germplasm accessions for yield characters recorded highest coefficient of variation was shown by wet weight of capsules (63.26%) followed by number of capsules (55.52%). Characters with highest genetic advance could be utilized for selection programmes as reported by earlier workers in cardamom (George et al. 1981; Radhakrishnan et al. 2006).

According to Korikanthimath et al. (1999a, b), there was a substantial positive association between the total number of tillers, bearing tillers, and panicles per plant and the number of capsules per plant. Selection based on these features can be used to increase cardamom because yield has a highly significant and positive link with both the number of capsules per plant and the length of the panicle. Gopal et al. (1990) reported similar association between traits. Backiyarani et al. (2002) established a highly significant and positive correlation between yield and number of panicles, panicle length, plant height, number of tillers as well as seed number and significant negative correlation between tiller number and weight of 100 capsules and concluded that bold capsule coupled with high recovery percentage is an ideal character for selection.

Length of panicle, total number of capsules per panicle, and wet weight of capsule per panicle showed positive correlation with dry capsule yield. Compact panicles with a greater number of capsules per node influence the yielding ability of clones (Korikanthimath and Mulge 2005). Senthil Kumar et al. (2018) reported that fresh weight of capsules (yield) per plant exhibited highly significant and positive correlation with number of capsules per plant (0.999) and panicle length (0.762) which indicated that these traits can be utilized for yield improvement. Jacob et al. (2020) while studying plant morphological traits associated with field resistance to cardamom thrips, significant association was found between panicle type, flower bract (persistent or nonpersistent), and attachment of the leaf sheath to pseudostem. Heryanto and Syukur (2021) studied variability among 100 open-pollinated progenies and observed highest variability for number of tillers followed by number of leaves. Highest positive correlation was observed between plant height and leaf length and observed no significant negative correlation between any traits. Cluster analysis showed 100 cardamom lines divided into three major groups with dissimilarity distance 12.

3.2.3 Inheritance and Genetic Analysis

In order to ascertain the mode of inheritance of resistance, Prasath et al. (2010) crossed two susceptible genotypes, namely CCS 1 and RR 1, with cardamom lines, NKE 9 and NKE 12, which are resistant to cardamom mosaic virus (CdMV). The findings showed that NKE 9 and NKE 12's CdMV resistance is controlled by genetics. Resistant F1 hybrids were produced by crossing susceptible and resistant genes. It was hypothesized by the disease reaction segregation pattern in the F2 and BC1 generations of the two crossings that two dominant complementary genes may be responsible for controlling CdMV resistance in NKE 9 and NKE 12. Overall, it was predicted that the resistance to CdMV is quantitative, dependent on gene dosage with totally dominant gene action, and possibly caused by two key causes.

Prasath and Venugopal (2002) carried out analysis of variance for combining ability effects of different characters showed that general (gca) as well as specific combining ability (sca) were highly significant for all the characters indicating that both additive and non-additive (dominance) gene action played important role in the expression of all the characters. Prasath and Venugopal (2004) evaluated 390 accessions of cardamom germplasm for 16 different characteristics. Nine characteristics exhibited a high degree of diversity. All of the genotypes were sorted into three clusters with varying numbers of genotypes after non-hierarchical Euclidean cluster analysis was performed. Accessions belonging to three distinct cultivar groups frequently form clusters together, indicating a potential degree of ancestry among the three groups.

Two clusters were seen by Hrideek et al. (2011) with a linkage distance of 0.99. Three genotypes made up the second cluster, while eight genotypes made up the first. The second cluster was made up of Panikulangara-1, Panikulangara-2, Njallani, Ela Rani-2, Vali Green Gold, Ela Rani-3, Ela Rani-1, and ICRI-2. The smaller cluster was made up of Palakkudi, PNS Vaigai, and Wonder cardamom. ICRI-2, on the other hand, created a distinct cluster at a 0.985 linkage distance. Ela Rani-3 and Vali Green Gold were found to have the closest genotypes. It was also discovered that Panikulangara-2 and Njallani were closely related.

Hrideek et al. (2015) factor analysis was carried out for 32 characters; the yield characters were grouped into two factors and both contributed 28–57% of cumulative variance and yield per plant and internodal length were lead characters. Based on multivariate cluster analysis, Akhila et al. (2017) categorized 117 FGB accessions into 5 clusters with an estimated height of 12. It was discovered that while clusters 4 and 5 included 8 and 20 accessions, respectively, clusters 1 and 2 included an identical number of accessions (44). It is interesting to note that cluster 3 was unique as it had only one accession (FGB 10). These findings suggested that the genotypes associated with clusters 4 and 5 might serve as the main targets of selection for genetic advancement.

No distinct clusters were observed based on Malabar, Mysore, and Vazhukka ecotypes. The fact that distinct clusters of accessions from the same ecotypes were formed further suggested the close ties between these ecotypes. The traits of number of capsules, wet weight, dry weight, plant height, and leaf breadth were all highest in cluster 4; the traits of number of bearing tillers, number of panicles, number of leaves, and leaf length were highest in cluster 5, according to Sharon et al. (2020). These cluster means were based on trait (morphological and yield parameters).

3.3 Cardamom Breeding

3.3.1 Flower Structure and Biology

Flowers are borne in panicles, and panicles emerge from the base of the tiller. Each tiller may have 2–4 panicles. Panicle emergence starts with onset of monsoon from 10 to 12 months old tillers. Growth pattern of panicles will vary between the biotypes, viz. prostrate in Malabar, erect in Mysore type, and semi-erect in case of *Vazhukka*. Each panicle possesses long cane like peduncle consisting of nodes and internodes. In each node, there will be scale leaf in the axil of which flowers are borne on cincinnus so the panicle will be branched (Pattanshetty and Prasad 1976; Parameswar 1973). Multiple branching or compound branching types are also reported in cardamom. In case of multiple branching types, there will be further branching (secondary and tertiary branches) of central peduncle (Prasath and Venugopal 2009). Such type of branching will be noticed at the lower portion of the main peduncle or in the top portion or it may be present throughput the peduncle.

Cardamom flowers are bisexual and zygomorphic. Cardamom inflorescence is called panicle and in the panicle flowers open in succession from base towards the apex. Flowers are white in colour and are 30–35 mm long. Central lip of the flower is streaked violet or blue and trimerous in condition. Labellum appears to be oval and indistinctly 3-lobed. Fertile stamen is one and is without connective appendages but prolonged into a short crest. Anthers are adnate to the filament, 2-lobed, 0.6–0.7 cm, and dehisce vertically. Inferior ovary, trilocular with axial placentation. Stigma is small and is above the anther. Prominent labellum, position of stigma above anthers, and presence of nectar glands favour insect pollination in cardamom. There is no variation in the flower morphology with respect to different biotypes or varieties (Madhusoodanan et al. 2002). Whereas there is a farmer's selection with completely white colour flower without violet tinge.

Pattanshetti and Prasad (1972) reported that maximum number of flowers open during early morning from 3.30 to 8.00 a.m. Dehiscence of anthers takes place immediately after anthesis, i.e. between 5.30 and 6.30 a.m. Belavadi et al. (1993, 1997, 1998), Parvathi et al. (1993), Belavadi and Parvathi (1998) have given detailed information about pollination ecology and biology of cardamom in Karnataka. It is reported that pollinator activity starts at 7.30 a.m. and continues up to 6.30 p.m., and peak foraging activity is between 11.00 a.m. and 1.00 p.m. Major pollinating agent is honeybees (*Apis cerana*). Number of times the flower visited by pollinating agents depends on the prevailing weather conditions, and it will be more during clear sunny day (120 times) followed by cloudy day (57 times) and rainy day (20 times).

Easy, simple, effective, and reliable technique for crossing was developed (Bhat et al. 2010) for producing cardamom hybrids. Well-developed and unopened flower buds were selected for emasculation. After emasculation, the flowers were enclosed in a bag. Flowers prepared for hybridization and pollen donors were also isolated by enclosing them with straw tubes. Pollination was carried out by collecting unopened flowers from the donor parent, which were also tagged and enclosed in straw tubes with a cap to avoid contamination. Artificial hand pollination was accomplished by direct transfer of pollen from dehisced anthers of fully opened flowers of the donor parent to the stigma between 09.00 and 11.00 h. For cross-pollination, either the straw or only the cap was removed. Pollen grains were scraped from the anther of the donor flower with a needle, collected on petri plates or paper bags and deposited on the stigma using a needle. After pollination, the straw tube was again covered with the cap for one more day to avoid further pollination by bees. After one month, the fruits from labelled flowers were counted, and the fruit setting percentage was determined, which ranged from 40 to 50 among the different crosses. Steps followed in emasculation and pollination in cardamom is depicted in Fig. 3.1.

3.3.2 Breeding Objectives

Crop improvement programme of any crop focuses on higher yield and resistance to biotic/abiotic stresses. In case of cardamom also, breeding objectives will be as follows.

1. Genotypes with high and sustainable yield
2. Evolving synchronized flowering genotypes for reducing harvesting cycle
3. Bold, green capsules with a greater number of seeds per capsule
4. Resistance to biotic stresses, viz. *katte*, rhizome rot, and thrips
5. Abiotic stress tolerance, viz. drought
6. High-quality genotypes with high percentage of oil, oleoresin, and α-terpenyl acetate content which imparts characteristic aroma and flavour to cardamom.

3.3.3 Breeding Methods

Amenability of cardamom for both sexual as well as asexual propagation gives good scope for crop improvement. Crop improvement techniques, viz. selection, heterosis breeding, mutation, and polyploidy breeding are used in cardamom.

3.3.3.1 Selection

Selection is one of the most common methods used in improvement of cardamom. Many of the improved and farmers varieties of cardamom are developed through selection. Cardamom is cross-pollinated crop so it will offer tremendous scope for increasing the productivity by selection of high yielding genotypes (Venugopal 1999). Korikanthimath et al. (1999a, b) evaluated 275 seedlings progenies of Cl-37 for selection of early and high yielding genotypes. Among the 275 genotypes, 49 entries were found to be high yielding. Yield varied from 325 to 7555 g wet capsules/clump. Entry no. 186 was found to be high yielding and early genotype. Korikanthimath et al. (2000) evaluated 16 cardamom genotypes for yield, recovery, and oil content. Based on yield AS 6/3 (632.5 kg/ha), Clone 779 (612.5 kg/ha) and S. 1272-5 (605 kg/ha) were selected as superior genotypes.

Twelve elite clones with local check were evaluated for yield by Korikanthimath et al. (2002), and it was reported that growth of plant, number of panicles, capsules, and wet capsule weight per plant will attribute to selection of superior clones with respect to total dry capsule yield. Compact panicle genotypes with a greater number of capsules per panicle and per node should be one of the selection criteria in identifying high productivity genotypes in cardamom (Korikanthimath and Mulge 2005).

Fig. 3.1 Emasculation and pollination in cardamom. **a** Selected flower bud for emasculation, **b** removal of sepals, **c** removal of petals, **d** stamens and stigma, **e** removal of stamens, **f** emasculated flower bud, **g** dehisced pollens, **h** artificial pollination

Venugopal et al. (2006) developed IISR Avinash variety with high yield (643 kg/ha), dark green capsules, and rhizome rot disease resistance through clonal selection. Prasath et al. (2009a, b) evaluated 14 genotypes of cardamom including 10 promising genotypes, one released variety Appangala 1 and three farmers' varieties. Among the genotypes studied, APG 250 and APG 416 performed better in terms of yield. Hrideek et al. (2011) evaluated ten farmers' varieties of cardamom along with released variety ICRI-2. Among the varieties evaluated, Ela Rani-2, Ela Rani-3, and Palakkudi were best performing in terms of growth and yield parameters. Akhila et al. (2017) carried out comparative analysis of overall performance of the four landraces and reported that Pappalu, Kalarikkal Bold, and Pulari performed significantly superior to Njallani Green Gold and ICRI-2.

Nadukeri et al. (2020a, b) evaluated 67 cardamom working collections for quantitative traits for two years and from pooled analysis concluded that KMRD-2, CSS-800, SKP-170, 26-16-D11, SEL-98, RR-1, and PDP-12 were found better with respect to morphological traits. Sharon et al. (2020) evaluated 117 field gene bank accessions for various phenotypic and yield characters. Among the 117 accessions, FGB 13 was selected as the superior genotype in terms of highest fresh weight of capsules (399.61 g) and with a greater number of capsules per plant (244.20).

3.3.3.2 Heterosis Breeding

Cardamom is amenable for both sexual as well as asexual propagation so hybridization will be very useful method of crop improvement. The *Vazhukka* type of cardamom is said to be the natural hybrid between Malabar type and Mysore type. In India only one species, i.e. *Elettaria* exists; hence, in cardamom crossing is confined to intra-specific levels (Madhusoodanan et al. 2002). As *Elettaria* is monotypic species, inter-specific hybridization is not possible, and in the intergeneric hybridization also no seed set was observed (Krishnamurthy et al. 1989). So only intra-specific crosses can be carried out in cardamom. Venugopal and Padmini (1999) observed good compatibility between eight Malabar cardamom genotypes crossed, and they opined that degree of compatibility varies with the parents selected for hybridization. Inter-varietal and inter-cultivar hybridizations were carried out to develop varieties/hybrids with desirable traits. Diallele crosses were carried out to combine the characters of early bearing, bold capsules, high yield, long panicle, leaf rot resistance, and multiple branching. Thirty cross-combinations were made, and the hybrid progenies were vigorous than the parental lines (Krishnamurthy et al. 1989).

Padmini et al. (2000a, b) studied heterosis for seedling characters in hybrid progenies. Among the 54 hybrids, nine hybrids exhibited significantly higher and positive heterobeltiosis and economic heterosis for plant height, leaf length, and leaf breadth except number of leaves per plant. Authors opined that these characters can be used while selecting superior hybrid progenies. Padmini et al. (2000a, b) evaluated hybrids, open-pollinated progenies, and inbreds and concluded that hybrids perform better than open-pollinated types and selfed parents indicating the hybrid vigour in these crosses which could be exploited. Open-pollinated progenies outperformed inbreds indicating the release of residual variability. Cardamom being cross-pollinated crop inbreeding depression is the major cause for reduced performance of inbreds.

Prasath et al. (2009a, b) crossed eight genetically diverse cardamom lines in a diallel fashion to study the hybrid performance and heterosis. The hybrid progenies were evaluated along with check variety IISR Avinash, and out of 56 hybrids, 10 hybrids out yielded IISR Avinash. Two hybrid progenies CCS 19 × NKE 19 and RR 19 × NKE 12 were identified for further evaluation trials. These two hybrids were having high yield and mosaic resistance. It was also reported that heterosis breeding helps to develop cardamom hybrids with high yield potential, desirable quality characters, and mosaic resistance by choosing appropriate parents. Akshitha et al. (2021) evaluated progenies of 23 intervarietal F_1 hybrids, and it was observed that among the hybrid progenies nine hybrid progenies were best performing in terms of yield. Among the nine hybrids, ICRI 4 × IISR Vijetha recorded highest fresh (1090 g) as well as dry yield (236 g) of capsules per plant.

3.3.3.3 Mutation Breeding

Physical mutagens, viz. X-rays and gamma rays, and chemical mutagens, viz. ethyl methane sulphonate and maleic hydrazide, were commonly used to induce mutation in cardamom. Large population of selfed and open-pollinated progenies of M1 plants were repeatedly inoculated with *katte* virus vector, and 12 plants were selected as tolerant to *katte* disease (Bavappa 1986). But so far there are no reports of mutants developed in cardamom.

3.3.3.4 Polyploidy Breeding

Sprouting seeds of cardamom were treated with 0.5% aqueous solution of colchicine to induce polyploids ($2n = 4x = 96$). These polyploids showed gigantism, increased layer of epidermal cells, thicker cuticle, and higher wax coating on the leaves (Sudharshan 1987, 1989).

3.3.4 Breeding for Biotic and Abiotic Stress

3.3.4.1 Biotic Stress

Cardamom is prone to infected by many fungal, bacterial, and viral diseases. Around 26 diseases have been reported which infect cardamom, and in that only few are found to be very severe which will cause severe crop loss (Thomas and Bhai 1995). Among the diseases, capsule rot, rhizome rot, leaf blight, leaf spot, and cardamom *katte* disease are the major ones. Cardamom is also infected by number of pests, and among them, major ones are thrips, shoot, and capsule borer. Venugopal et al. (2006) surveyed rhizome rot disease hotspots (Madenadu, Kandanakolly and Halery in Kodagu District, Karnataka), and observations were recorded on 51,000 plants. Out of which four plants free from natural infection of rhizome rot for about 4–5 years and having good yield were selected.

Disease escapes along with germplasm accessions were screened artificially and found that with injury disease index was 8–56% and without injury it was 4–38%. Out of all the entries under trial, RR-1 exhibited 10% disease index through pin pricking and without injury only 4% disease index was observed. This entry (RR-1) was evaluated in coordinated varietal trial under All India Coordinated Research Project on Spices and was found superior in terms of yield (643 kg/ha), dark green capsules, and resistance to rhizome rot disease. This led to the development and release of RR-1 as IISR Avinash variety of cardamom.

Evaluation and characterization of germplasm also pave way for identification of superior genotypes in terms of yield, quality, and biotic and abiotic resistance. In a similar attempt by Senthil Kumar et al. (2018) identified rhizome rot resistant and leaf blight disease-resistant genotypes. From the study, genotypes with dual resistance against leaf blight and rhizome rot were identified. Genotypes IC 584083, IC 584087, IC 547224, IC 547188, IC 584081, IC 584084, IC 547185, IC 547183, IC 547226, IC 547202, IC 547203, IC 547204, and IC 547210 exhibited resistance to leaf blight and rhizome rot. Screening of 180 germplasm accessions of cardamom resulted in shortlisting 8 accessions, viz. IC 349370, IC 349606, IC 349588, IC 349363, IC 349362, IC 349364, IC 349455, and IC 547167 with resistance to cardamom thrips. Average capsule damage in these resistant accessions ranged from 13.7 to 16.7% (Jacob et al. 2020).

From the screening of 117 field gene bank accessions of cardamom at Appangala accessions, FGB 35 and FGB 15 were identified as resistant and highly resistant to leaf blight and rhizome rot, respectively (Sharon et al. 2020). Akshitha et al. (2021) screened high yielding progenies of 23 inter-varietal crosses for rhizome rot resistance. It was found that hybrid progenies, viz. PV 2 × IISR Vijetha (progeny no. 1), Mudigere 3 × IISR Vijetha (progeny no.

5), ICRI 4 × IISR Avinash (progeny no. 2), ICRI 4 × IISR Vijetha (progeny no. 14), Mudigere 1 × IISR Vijetha (progeny no. 1), and Mudigere 1 × IISR Vijetha (progeny no. 2) did not show any symptoms of rhizome rot during challenge inoculation under pot culture conditions. These hybrid progenies are shortlisted as resistant to rhizome rot disease.

134 disease escapes from hotspots of virus infection in South India were gathered, and they were screened in greenhouses, sick plots, and hotspots to identify resistant sources to cardamom mosaic virus, which causes katte disease of cardamom. Promising collections were tested against spontaneous infection and in four hotspots, confirming their resistant character (Venugopal 1999). Later, one of this high yielding selection was released as IISR Vijetha. Out of all the hybrid combinations, hybrid IC 547167 was discovered to have a rich concentration of α terpinyl acetate, making it similar to Appangala 1, and to be relatively resistant to the cardamom mosaic virus under artificial conditions. In 2014, the genotype IC 547167 was released as Appangala 2 for Karnataka and the surrounding districts of Kerala due to its high production and tolerance to the cardamom mosaic virus in field circumstances. Hybridization between Appangala 1 × NKE 19 produced the variety Appangala 2 (IC 547167) (Senthil Kumar et al. 2017).

Field screening of 180 cardamom germplasm accessions for three years at Appangala, Karnataka, resulted in identification of eight accessions resistant to cardamom thrips (Jacob et al. 2020).

3.3.4.2 Abiotic Stress

Moisture stress is one of the important factors which hinders crop growth most of the crop plants. Cardamom is also one such crop which needs proper irrigation during the summer months. During December-May cardamom experiences moisture stress (Hegde and Korikanthimath 1996). Gurumurthy et al. (1996) observed significant yield differences among 12 cardamom clones evaluated under regulated natural shade and moisture levels. Among the 12 clones evaluated P3 and P6 were found to be superior.

For characterization of cardamom germplasm for moisture stress, morphological traits such as maintenance of leaf area, plant height, and biomass along with physiological parameters, viz. relative water content and membrane stability can be used (Ankegowda and Krishnamurthy 1998).

Ankegowda and Krishnamurthy (2008) evaluated six cardamom genotypes with high biomass and good yield for moisture stress tolerance. Among them, Mysore 2, Hybrid 36, and Green Gold were found to have better adaptability under moisture stress conditions due to maintenance of more leaf area, more number of tillers, and total dry weight. Alagupalamuthirsolai et al. (2020) developed rapid screening method for drought tolerance in cardamom using polyethylene glycol 6000. Seedlings of Appangala 1 variety cardamom were treated with four different concentrations of PEG (5, 10, 15, and 20%). PEG-6000 concentration above 15% has reduced percent seedling survival almost by 50%. At 15% PEG 6000, there observed increase in proline, phenolic content, enzyme scavenging activity and decrease in chlorophyll fluorescence. By this it was concluded that PEG 15% is the ideal concentration for screening of cardamom seedlings for moisture stress.

3.3.5 Varietal Status

3.3.5.1 Improved Varieties

Through the selection of superior plants with desirable characteristics of higher yield and superior capsule traits from land race populations, the improved varieties that have been identified thus far have evolved. This achievement has been attributed to the crop's extremely heterozygous character and the capacity to replicate selections through clonal propagations. In terms of yield and capsule characteristics, the enhanced selections are significantly better than the local clones. Table 3.1 summarizes the key characteristics of the current crop of cultivars.

Table 3.1 Salient features of improved varieties of cardamom

Variety/hybrid	Type	Average yield (kg/ha)	Important features
ICAR-Indian Institute of Spices Research, Regional Station, Appangala, Madikeri, Karnataka			
Appangala 1	Malabar	745	Suitable for intensive cultivation both under monocrop and mixed crop conditions. Early maturing variety, highly adaptive and produces 89% bold capsules. (Potential yield 1322 kg dry capsules/ha) Suitable for cultivation in all cardamom growing tracts of Karnataka and Wayanad of Kerala
IISR Avinash	Malabar	848	Rhizome rot resistant, high yielder, and suitable for planting in valleys. Produces 51% bold, dark green capsules. (Potential yield 1483 kg dry capsules/ha) Suited for hotspots of rhizome rot-prone areas
IISR Vijetha	Malabar	643	*Cardamom mosaic virus*-resistant (*katte*) selection, recommended for moderate to high-shaded mosaic disease-prone areas. (Potential yield 979 kg dry capsules per ha) Suitable for Kodagu, Hassan, Chikkamagalur, and North Wayanad Suitable for *katte*-prone areas. Adapted to moderate rainfall and moderate to high shade areas
Appangala 2	Malabar	927	Hybrid, high yielding, and resistant to *cardamom mosaic virus/katte* Suitable for cultivation in all cardamom growing tracts of Karnataka and Wayanad of Kerala
Indian Cardamom Research Institute (ICRI), Myladumpara, Kerala			
ICRI 1	Malabar	656	Early maturing, dark green bold capsules. Suitable to South zone of Idukki, Kerala
ICRI 2	Mysore	766	Tolerant to *Azhukal* disease. Suitable for cultivation in Vandanmedu and Nelliyampathy of Kerala and Anamalai, Meghamalai of Tamil Nadu
ICRI Regional Station, Sakaleshapura, Karnataka			
ICRI 3	Malabar	599	Tolerant to rhizome rot. Cardamom growing tracts of Karnataka
Indian Cardamom Research Institute (ICRI), Myladumpara, Kerala			
ICRI 4	Malabar	961	Suitable for low rainfall regions. Adapted to lower Palani hills
ICRI 5	Malabar	1543	Hybrid with high volatile oil (7.13%) content. Suitable to Kerala and parts of Tamil Nadu
ICRI 6	Malabar	1900	Moderately tolerant to rot, thrips, borer, and drought. Suitable to Kerala and parts of Tamil Nadu
ICRI 7	*Vazhuka*		Deep green bold capsules, moderately tolerant to azhukal disease
ICRI 8	Malabar	706	Ovoid bold, pale green capsules, higher recovery of bold capsules with 70.1% of the capsules produced is above 7.5 mm in diameter
Zonal Agricultural and Horticultural Research Station (University of Agricultural and Horticultural Sciences), Mudigere, Karnataka			
Mudigere 1	Malabar	275	Erect and compact plant, pale green, oval bold capsules, suitable for high density planting, moderately tolerant to thrips, hairy caterpillar, and white grubs, and dry recovery 20.0% Suitable for cultivation in traditional cardamom growing Malanad areas of Karnataka, under rainfed conditions

(continued)

Table 3.1 (continued)

Variety/hybrid	Type	Average yield (kg/ha)	Important features
Mudigere 2	Malabar	475	Early maturing, suitable for high density planting, round/oval and bold capsules, oil 8.0%, 1,8-cineole 45.0%, α-terpenyl acetate 38.0%. Suitable for cultivation in cardamom growing tracts of Karnataka
Mudigere 3	Malabar	900–1100	Leaves non-pubescent, capsules oval-oblong, resistant to leaf spot disease, least incidence against capsule rot and borer. Suitable for cultivation in cardamom growing tracts of Karnataka
Cardamom Research Station (Kerala Agricultural University), Pampadumpara, Kerala			
PV 1	Malabar	500	An early maturing type, short panicle, elongated slightly ribbed light green capsules, essential oil 6.8%, 1,8-cineole 33.0%, α-terpenyl acetate 46.0%, dry recovery 19.9%. Suitable for cultivation in all cardamom growing tracts of Kerala and parts of Tamil Nadu
PV 2	*Vazhukka*	982	Early maturing, unbranched lengthy panicle, long bold capsules, high dry recovery percentage (23.8%), and essential oil 10.45% field tolerant to stem borer and thrips. Suitable for elevation range of 1000–1200 m above MSL. Suitable for cultivation in Cardamom Hill Reserves of Idukki
PV 3	Malabar		Ovoid capsule, unbranched panicle, relatively tolerant to capsule rot, rhizome rot, and thrips

3.3.5.2 Farmers Selections

There are many high yielders, traditional cultivars or landraces in cardamom that have been selected and developed by farmers through years of cultivation and selection. Some of the important selections are given in Table 3.2.

3.3.6 Constraints of Cardamom Breeding

Development of improved cultivars in cardamom may require many breeding cycles and dozens of years because of semi-perennial nature. The monotypic genus status of cardamom in India offers little space for concerted breeding efforts. Also, compatibility barrier between Elettaria and other closely related genera affects introgression of useful genes through wide hybridization. The lack of genomic information in cardamom hinders acceleration in breeding using marker-assisted breeding.

3.4 Conclusion and Future Prospectus

Cardamom is an important spice crop, which has been prominent in world trade since ancient times. During the past two decades, world cardamom production and demand have increased greatly. Considering huge genetic diversity available in the crop, selective breeding and cross-breeding strategies may be intensified in developing high yielding and disease, pest-tolerant genotypes. Development of mapping population and extensive genotyping and phenotyping is very crucial to identify genomic regions controlling yield and quality contributing characters. The whole genome and transcriptome approaches based on NGS have shown to be more effective than earlier techniques and are quickly taking the lead in NGS research on spices. A method known as functional genomics/metabolomics aids in the clarification of the biochemical routes and important genes

Table 3.2 Characteristic features of farmers selections in cardamom

Variety	Name and address of farmer	Type	Salient features
Njallani green gold	Mr. Sebastin Joseph and Mr. Rejimon Joseph Njallaniyil House P. O. Kattappana Dist. Idukki Kerala	*Vazhukka*	High yielding variety. 80% of the capsules will be >7 mm
Thiruthali	Mr. T. P. Joseph Idukki, Kerala	*Vazhukka*	Medium bold, parrot green colour capsules. Good yielder under rainfed conditions. Suited for areas with less shade (< 40%). 4–6 kg dry yield per plant
Wonder cardamom	Mr. Sabu Varghese Idukki, Kerala	*Vazhukka*	Panicles are branched and very long ranging from 1.5 to 2.0 m with short internodes. Capsules are extra bold in size with deep green colour. 3–4 kg dry yield/plant. This variety can be intercropped in rubber plantation
Panikulangara green bold 1	Mr. Joy Peter Idukki, Kerala	*Vazhukka*	Capsules retain attractive green colour after drying. Can be cultivated in less shaded (40%) areas
Panikulangara green bold 2	Mr. Joy Peter Idukki, Kerala	*Vazhukka*	Resistant to *Azhukkal* disease
White flower	Mr. K. J. Baby Puliyanmala, Idukki, Kerala	*Vazhukka*	Flowers will be pure white in colour. 4–5 kg dry cardamom per plant. Essential oil content 8.9%
PNS Vaigai	Mr. P. N. Surulivel Theni, Tamil Nadu	Mysore	High and consistent yielder—900–1400 kg/acre. 60–70% of the produce is 7 mm and above
Elarajan	Mr. K. J. Benny Kerala	*Vazhukka*	80–90% of capsules will be 8–9 mm. High yielding 2500 kg/ha
Pappalu	Mr. K. V. Paulose Idukki, Kerala	*Vazhukka*	Average yield 2000–2500 kg/ha. 25% dry recovery. Tolerant to thrips, stem borer, root grub, and capsule rot. Long panicle (130–147 cm), ovoid capsules
Arjun	Mr. Menuwin Thomas Idukki, Kerala	*Vazhukka*	Suitable to higher elevation (3000 ft from sea level). 2000–3000 kg dry yield/ha. Resistant to thrips and borer
Pachaikkai	Mr. L. Ramaiah and Mr. R. Murugan Theni, Tamil Nadu	*Vazhukka*	High yielding 2000 kg dry yield/ha. Bright green coloured bold capsules. Drought tolerant and most suitable for lower elevation areas of Tamil Nadu
Thadathil gold	Mr. C. M. Mathew Idukki, Kerala	*Vazhukka*	Oviod and pale green capsules. High yielding variety

that underlie particular cardamom nutraceutical substances. A concerted effort is required in classical breeding techniques along with recent biotechnological intervention to facilitate the greater improvement of cardamom for yield and quality.

References

Akhila PR, Radhakrishnan VV, Pradipkumar K et al (2017) Assessment of variability and performance of new landraces of small cardamom (*Elettaria cardamomum* Maton). J Plant Crops 45(1):43–48

Akshitha HJ, Peeran MF, Aravind S et al (2021) Evaluation of small cardamom (*Elettaria cardamomum* Maton.) hybrids for yield and rhizome rot resistance. In: Biju CN, Alagupalamuthirsolai M, Senthil Kumar CM, Dinesh R, Prasath D, Ramakrishnan Nair R, Eapen SJ (eds) e-Abstracts-SYMSAC X 2021. International symposium on spices as flavours, fragrances & functional foods, 09–12 Feb 2021. Indian Society for Spices, Kozhikode, Kerala, India

Alagupalamuthirsolai M, Ankegowda SJ, Peeran MF et al (2020) Establishment of a rapid screening method for drought tolerance of small cardamom by using polyethylene glycol-6000. Int J Chem Stud 8(2):1750–1755. https://doi.org/10.22271/chemi.2020.v8.i2aa.9013

Anisha CS, Mary Mathew K, Sasidharan S et al (2020) Diversity analysis of released varieties of Indian cardamom using ISSR markers reveal narrowing genetic base. Indian J Biotechnol 19:311–322

Anjali N, Nadiya F, Shefeek S, Sabu KK (2016a) Genome size estimation using flow cytometry in *Elettaria cardamomum* Maton. Indian Forester 142(9):878–881

Anjali N, Ganga KM, Nadiya F et al (2016b) Intraspecific variations in cardamom (*Elettaria cardamomum* Maton): assessment of genomic diversity by flow cytometry, cytological studies and ISSR analysis. SpringerPlus 5:1560. https://doi.org/10.1186/s40064-016-3226-x

Ankegowda SJ, Krishnamurthy KS (1998) Effect of moisture stress on growth of cardamom (*Elettaria cardamomum* Maton). In: Sadanandan AK, Krishnamurthy KS, Kandamman K, Karikanthianath VS (eds) Water and nutrient management for sustainable production and quality of spices. Indian Society for Spices, Calicut, pp 158–161

Ankegowda SJ, Krishnamurthy KS (2008) Evaluation of small cardamom accessions for moisture stress. J Spices Aromat Crops 17(2):172–176

Backiyarani S, Kurian S, Josephrajkumar A et al (2002) Evaluation of high yielding accessions of small cardamom (*Elettaria cardamomum* Maton) for suitability in the high ranges of Idukki District. J Spices Aromat Crops 11(2):93–96

Bavappa KVA (1986) Research at CPCRI. Technical bulletin. CPCRI, Kasaragod, India, p 34

Belavadi VV, Parvathi C (1998) Estimation of honeybee colonies required for effective pollination in cardamom. In: Proceedings of national symposium diversity of social insects & other arthropods and the functioning ecosystems, III congress of IUSSI, Mudigere, p 30

Belavadi VV, Chandrappa HM, Shadakshari YG et al (1993) Isolation distance for seed gardens of cardamom (*Elettaria cardamomum* Maton). In: Veeresh GK, Umashanker R, Ganeshan KN (eds) Pollination in tropics. IUSSI, Bangalore, pp 241–243

Belavadi VV, Venkateshalu, Vivek HR (1997) Significance of style in cardamom corolla tubes for honeybee pollination. Curr Sci 73:287–290

Belavadi VV, Parvathi C, Raju B (1998) Optimal foraging by honeybees on cardamom. In: Proceedings of XIII plantation crops symposium, Coimbatore, 16–18 Dec 1998

Bhat SS, Thomas J, Madhusoodanan KJ et al (2010) A novel hybridization technique in cardamom (*Elettaria cardamomum* Maton). J Plant Crops 38(1):75–77

Chakravarti AK (1948) Multiplication of chromosome numbers in relation to speciation in Zingiberaceae. Sci Cult 14:137–140

Chandrasekhar R, Sampath Kumar R (1986) Karyological highlights on two cultivars of cardamom (*E. cardamomum* Maton). J Cytol Genet 21:90–96

George KV, Dandin SB, Madhusoodanan KJ et al (1981) Natural variations in the yield parameters of cardamom (*Elettaria cardamomum* Maton). In: Visveshwara S (ed) Proceedings of PLACROSYM 1981. Indian Society for Plantation Crops, Kasaragod, India, pp 216–231

Gopal R, Chandraswamy DW, Nayar NK (1990) Correlation and path analysis in cardamom. Indian J Agric Sci 60:240–242

Gopal R, Chandraswamy DW, Nayar NK (1992) Genetic basis of yield and yield components in cardamom. J Plant Crop 20(Suppl):230–232

Gregory PJ (1936) The floral morphology and cytology of *Elettaria cardamomum* Maton. J Linn Soc 50:363–391

Gurumurthy BR, Chandrappa HM, Parvathi C et al (1996) Screening of cardamom clones for drought tolerance. In: Satheesan KV (ed) Proceedings of the national seminar on drought management in plantation crops 1996, Kottayam, Kerala, pp 25–29

Hegde R, Korikanthimath VS (1996) Agronomic approaches for drought management in cardamom. In: Satheesan KV (ed) Proceedings of the national seminar on drought management in plantation crops 1996, Kottayam, Kerala, pp 75–79

Heryanto R, Syukur C (2021) Study of genetic diversity and relationship of 100 cardamom (*Elettaria cardamomum*) lines based on morphological characters. IOP Conf Ser Earth Environ Sci 752:012046

Hrideek TK, Radhakrishnan VV, Mohanan KV et al (2011) Genetic divergence in some elite landraces of small cardamom (*Elettaria cardamomum* Maton). J Plant Crop 39:201–202

Hrideek TK, Radhakrishnan VV, Kuruvilla KM et al (2015) A study on variability of elite landraces of small cardamom (*Elettaria cardamomum* Maton). Indian J Plant Genet Resour 28(3):311–316

Jacob TK, Senthil Kumar CM, Devasahayam S et al (2020) Plant morphological traits associated with field resistance to cardamom thrips (*Sciothrips cardamomi*) in cardamom (*Elettaria cardamomum*). Ann Appl Biol 177:143–151. https://doi.org/10.1111/aab.12592

Korikanthimath VS, Mulge R (2005) Studies on panicle and capsule characters in elite clones of cardamom. Indian J Hortic 62(4):367–369

Korikanthimath VS, Mathew MJ, Mulge R et al (1999a) Studies on selection of elite lines of CI-37 seedling progenies of cardamom (*Elettaria cardamomum* Maton). J Spices Aromat Crops 8(1):57–61

Korikanthimath VS, Mulge R, Hegde R et al (1999b) Correlation between yield and yield parameters in cardamom (*Elettaria cardamomum*). J Med Aromat Plants 21:700–701

Korikanthimath VS, Prasath D, Mohammed Sayed AA (2000) Evaluation of cardamom accessions in relation to yield, recovery and oil content. Indian J Hortic 57(3):273–275

Korikanthimath VS, Mulge R, Rajendra H et al (2002) Evaluation of elite lines of cardamom (*Elettaria cardamomum*) for yield and yield parameters. Indian J Hortic 59(3):326–330

Krishnamurthy K, Khan MM, Avadhani KK et al (1989) Three decades of cardamom research at R.R.S., Mudigere. Technical bulletin. University of Agricultural Sciences, Bangalore, India

Madhusoodanan KJ, Pradip Kumar K, Ravindran PN (2002) Botany, crop improvement and biotechnology of cardamom. In: Ravindran PN, Madhusoodanan KJ (eds) Cardamom: the genus *Elettaria*. Taylor & Francis, London, pp 11–68

Nadukeri S, Umesha K, Naik MK et al (2020a) Evaluation of working collections of cardamom (*Elettaria cardamomum* Maton) for morphological attributes under hill zone of Karnataka. Int J Chem Stud 8(6):2215–2219

Nadukeri S, Umesha K, Naik MK et al (2020b) Variability studies in small cardamom (*Elettaria cardamomum*) genotypes for growth characters. J Pharmacogn Phytochem 9(6):1346–1349

Padmini K, Venugopal MN, Ankegowda SJ (2000a) Heterosis for seedling characters in cardamom (*Elettaria cardamomum* Maton). J Spices Aromat Crops 9(2):149–153

Padmini K, Venugopal MN, Sasikurnar B (2000b) Performance of hybrids, open pollinated progenies and inbreds of cardamom (*Elettaria cardamomum* Maton) under nursery conditions. Indian J Agric Sci 70(8):550–551

Parameswar NS (1973) Floral biology of cardamom (*Elettaria cardamomum* Maton). Mysore J Agric Sci 7:205–213

Parvathi C, Shadakshari YG, Belavadi VV et al (1993) Foraging behaviour of honeybees on cardamom (*Elettaria cardamomum* Maton). In: Veeresh GK, Umashanker R, Ganesan KN (eds) Pollination in tropics. IUSSI, Bangalore, pp 99–103

Pattanshetti HV, Prasad ABN (1972) Blossom biology, pollination, and fruit set in cardamom (*Elettaria cardamomum* Maton). In: Third international symposium on subtropical and tropic, pp 262–268

Pattanshetty HV, Prasad ABN (1976) Blossom biology, pollination, and fruit set incardamom. In: Chadha KL (ed) Proceedings of international symposium on subtropical and tropical horticulture, vol 1. Today and Tomorrow's Publishers, New Delhi, pp 262–268

Prasath D, Venugopal MN (2002) Genetic analysis for quantitative characters in cardamom (*Elettaria cardamomum* Maton). Indian J Genet Plant Breed 62:179–180

Prasath D, Venugopal MN (2004) Genetic diversity and conservation of cardamom (*Elettaria cardamomum* Maton) in India. Plant Genet Resour Newslett 138:55–60

Prasath D, Venugopal MN (2009) Compound inflorescence cardamom (*Elettaria cardamomum* Maton) in India. Genet Resour Crop Evol 56:749–753

Prasath D, Venugopal MN, Korikanthimath VS (2001) Genetic variability in cardamom (*Elettaria cardamomum* Maton.). Indian J Plant Genet Resour 14(2):217–218

Prasath D, Venugopal MN, Senthil Kumar R (2009a) Evaluation of high yielding cardamom selection in Karnataka for yield and quality. Indian J Agric Sci 79(3):207–209

Prasath D, Venugopal MN, Senthilkumar R et al (2009b) Hybrid performance for yield and yield components in cardamom (*Elettaria cardamom* Maton). Euphytica 168:49–60

Prasath D, Venugopal MN, Parthasarathy VA (2010) Inheritance of cardamom mosaic virus (CdMV) resistance in cardamom. Sci Hortic 125:539–541

Radhakrishnan VV, Mohanan KV, Priya PM (2006) Genetic divergence in cardamom (*Elettaria cardamomum* Maton). J Plant Crops 34:149–151

Senthil Kumar R, Prasath D, Aravind S et al (2017) Evaluation of elite cardamom (*Elettaria cardamomum*) genotypes for yield, quality, and resistance to cardamom mosaic virus. J Plant Crops 45(2):129–134

Senthil Kumar R, Sharon A, Biju CN et al (2018) Characterization of genotypes of small cardamom (*Elettaria cardamomum* Maton) for yield parameters and disease resistance. J Plant Crops 46(2):143–146

Sharma AK, Bhattacharyya NK (1959) Cytology of several members of Zingiberaceae and the study of the inconsistency of their chromosome complements. La Cellule 59:297–346

Sharon A, Biju CN, Ankegowda SJ et al (2020) Screening small cardamom (*Elettaria cardamomum* Maton) field gene bank accessions for phenotypic characters, yield potential and disease resistance. J Spices Aromat Crops 29(2):105–112

Sudharshan MR (1987) Colchicine—induced tetraploids in cardamom (*Elettaria cardamomum* Maton). Curr Sci 56(1):36–37

Sudharshan MR (1989) Induced polyploids in cardamom. J Plant Crops 16:365–369

Thomas J, Bhai RS (1995) Fungal and bacterial diseases of cardamom and their management. J Spices Aromat Crops 4(1):24–31

Venugopal MN (1999) Natural disease escapes as source ofresistance against cardamom mosaic virus causing *katte* disease of cardamom (*Elettaria cardamomum* Maton). J Spices Aromat Crops 8:145–151

Venugopal MN, Padmini K (1999) Studies on crossability between elite types of cardamom (*Elettaria cardamomum* Maton). J Spices Aromat Crops 8(2):185–187

Venugopal MN, Prasath D, Mulge R (2006) IISR Avinash—a rhizome rot resistant and high yielding variety of cardamom (*Elettaria cardamomum* Maton). J Spices Aromat Crops 15(1):14–18

4 Micropropagation Methods in Small Cardamom

K. Mary Mathew, Y. S. Rao, H. N. Anjana, S. D. Soumya, and K. K. Sabu

Abstract

Cardamom (*Elettaria cardamomum* Maton) belongs to the family Zingiberaceae and is believed to be originated from the Western Ghats of Southern India. It is a cross-pollinated crop and therefore a lot of variability exists in its natural seed populations. Micropropagation techniques were developed for cardamom to obtain uniform plants in large numbers. The protocol has been transferred to several biotech firms in public and private sectors all over India for mass multiplication of elite varieties of cardamom. It is worthwhile to note that cardamom rhizomes cultured in MS medium with growth regulators produce friable callus for cell suspension culture, enhancing downstream studies.

Keywords

Cardamom · Micropropagation · Tissue culture · Cytokinins · Auxin

K. Mary Mathew (✉)
Indian Cardamom Research Institute—ICRI, Spices Board, GoI, Kochi (Retired), Lizlynne, Kuriachira, Thrissur, Kerala 680006, India
e-mail: marykaravat@gmail.com

Y. S. Rao
Indian Cardamom Research Institute ICRI, Spices Board, GoI, Kochi (Retired), Shalom, Pulse Nagar, Eroor North, Kochi 682306, India

H. N. Anjana · S. D. Soumya · K. K. Sabu
Biotechnology and Bioinformatics Division, Jawaharlal Nehru Tropical Botanic Garden and Research Institute (JNTBGRI), Palode, Thiruvananthapuram 695562, India

4.1 Introduction

Micropropagation is a technique that facilitates the efficient and extensive replication of genetically identical plant specimens. This process enables the rejuvenation of aged cultivars and the prompt regeneration of novel cultivars that exhibit resistance to both biotic and abiotic stresses. Additionally, micropropagation can persuade somaclonal variations, which contribute to the emergence of new traits. The technique offers benefits in addressing seed dormancy in recalcitrant plants and crops with limited availability of planting material. Additionally, it facilitates the cultivation of cells for the production of valuable plant products and secondary metabolites in vitro, thereby reducing the time and effort required. The need for tissue culture in cardamom has witnessed a notable surge due to its ability to generate virus-free plants. And it also reduces the duration of time required for establishment and commercial production by doing away with the demand for conventional propagation techniques like seeds or suckers.

K. K. Sabu and C. Kole (eds.), *The Cardamom Genome*, Compendium of Plant Genomes,
https://doi.org/10.1007/978-3-031-71829-8_4

Cardamom (*Elettaria cardamomum* Maton) is a tall, perennial, ornamental, herbaceous, rhizomatous monocotyledonous shrub of the family Zingiberaceae, originating from the Western Ghats of Southern India. India is the second-largest producer of cardamom in the world, behind Guatemala. In 2021–22 and 2022–23, the country produced about 47,803 MT of cardamom, of which 18,933 MT were exported, bringing in 294 million US dollars (Spices Board India 2023). The plant is valued for its dried fruits (capsules) containing aromatic seeds, which possess a strong, unique taste, resinous fragrance, and soothing aroma along with a cooling effect similar to mint. Cardamom has been used as both a spice and medicine since the fourth century, as it possesses carminative, stomachic, and anti-microbial properties (Chempakam and Sindhu 2008; Tyagi et al. 2009).

Owing to its highly outcrossing nature, seed propagation results in high genetic variability in the crop. According to Purseglove (1975), cardamom seed germination is frequently poor, erratic, and some seeds may even take more than a year to germinate. Tiller multiplication has been used for vegetative proliferation; however, this method is laborious and slow. Additionally, two dangerous diseases that make planting and conserving cardamom risky and challenging in the field genebank are cardamom mosaic virus (Katte disease) and *Pythium vexans* (rhizome rot), which pose a threat to vegetatively propagated crops (Purseglove et al. 1981). Infestations of entire cardamom plantations with "Katte" or mosaic disease are a major threat faced by growers. Due to its perennial nature and susceptibility to many environmental fluctuations, maintenance of cardamom germplasm in the field genebank is also labor-intensive. Therefore, in vitro conservation is considered to be a safe alternative strategy for conserving the germplasm of such crops. More precisely, tissue culture techniques allow for the easy maintenance and propagation of elite clones free from viruses and pathogens.

Cardamom is a cross-pollinated crop and therefore a lot of variability exists in its natural seed populations. Micropropagation techniques were developed for cardamom in the early 1980s to obtain uniform plants in large numbers for realizing consistency in yield and other qualitative traits. The earliest tissue culture propagation method in cardamom was from the Indian Institute of Horticultural Research (IIHR), Bangalore (Rao et al. 1982). Before long, Nadgauda et al. (1983) in collaboration with Indian Cardamom Research Institute (ICRI, Spices Board) developed a protocol for clonal multiplication of cardamom at NCL (National Chemical Laboratory) Pune, using rhizome buds. This protocol paved way for the first commercial production of plantlets through Tissue Culture (TC) and 'Tissue culture-small cardamom product plan' jointly sponsored by Spice Board and Department of Biotechnology (DBT), New Delhi in 1990s, a large-scale field evaluation trial of TC plantlets of cardamom. This was the first ever large-scale demonstration of tissue culture planting of a spice crop in farmers' fields in India, in more than 100 ha with 50 units in Kerala, 35 in Karnataka, and 15 in Tamil Nadu, each unit comprising 1 ha (Spices Board 1996). Around 25–37% increase in yield was recorded in tissue culture plantlets as against their seedling counterparts. However, Sudharsan et al. (1997) reported an overall variability of 4.5% in tissue-cultured plants as against 3.0% in open-pollinated seedling progeny for a given set of characters and suggested that variations were due to adventitious bud formation during micropropagation, genetic instability of adventitious meristem, and tissue culture-induced disorganization of meristems. The protocol was yet again revised and scaled up (Rao et al. 2000), and transferred to several biotech firms in public and private sectors all over India.

4.2 Micropropagation of Cardamom

Cardamom can be propagated by seeds as well as by suckers. Micropropagation techniques have been utilized for rapid and extensive propagation of disease-free, elite clones (Reghunath and Bajaj 1992). Cardamom is viable to tissue culture (Fig. 4.1), but there are only limited reports on its in vitro propagation using rhizomes (Babu et al. 1997), vegetative buds (Nadgauda et al. 1983; Reghunath and Gopalakrishnan 1991), somatic embryos (Manohari et al. 2008), callus (Rao et al. 1982; Priyadarshan and Zachariah 1986), immature floral buds (Kumar et al. 1985), shoot tips, and nodal explants (Tyagi et al. 2009).

Fig. 4.1 Micropropagated cardamom, 6 months old maintained at JNTBGRI, Palode

4.3 Effect of Growth Hormones on Shoot Proliferation

4.3.1 Effect of Cytokinins

An ideal medium for in vitro conservation should ensure higher survival rate, regular rooting, and slow development. The choice of an appropriate explant and composition of culture medium are crucial factors in any in vitro propagation procedure as they influence the growth of both the shoots and the roots. Proliferation rate, proportion of regenerating explants, and plant shoot length are all strongly impacted by the growth hormone composition of the culture medium. The ability of high concentrations of cytokinins to induce axillary branching is documented by many research groups (George et al. 2008). It has been observed that in medium enriched with cytokinin, the majority of herbaceous plants develop strong, well-formed shoots (Debergh and Zimmerman 1991).

Malhotra et al. (2021) reported that better rates of shoot multiplication and healthy growth were observed in cultures of cardamom on medium supplemented with BAP (6-Benzylaminopurine) alone or in combination with kinetin, conversely, reduced growth was observed in shoots cultured on medium enriched with NAA (Naphthalene acetic acid) alone or in combination with BAP. The maximum shoot number per explant (5.83) was observed on medium supplemented with 2.32 μM and 4.44 μM BAP. The longest shoots (6 cm) were seen on medium containing 2.32 μM Kn and 0.44 μM BAP and the shortest (2.5 cm) were seen on medium containing 5.37 μM NAA and 8.88 μM BAP.

Prior research has documented the beneficial role of BAP in the development of numerous shoots in cardamom tissue culture. In one of the earliest reports, precocious shoot growth was observed in shoot primordia cultivated on MS medium supplemented with BAP, Kn, and coconut milk, according to Nadgauda et al. (1983). In cardamom cultures, Reghunath and Bajaj (1992) found shoot differentiation on medium containing combinations of BAP and Kn, especially 3.0 mg L^{-1} BAP and 0.5 mg L^{-1} Kn as well as 2.0 mg L^{-1} BAP and 1.0 mg L^{-1} Kn. Ravindran and Babu (2002) have documented comparable findings regarding the function of BAP in the development of plantlets derived from cardamom microspores. Tyagi et al. (2009) also reported a high shoot proliferation rate on MS medium supplemented with 5 μM BAP.

4.3.2 Effect of Auxin

In cardamom, it was reported that NAA stimulates the development of callus. Reghunath and Bajaj (1992) reported that the addition of 4.0 mg L^{-1} NAA and 1.0 mg L^{-1} BAP to the growth medium led to a stimulating impact on callus formation in cardamom. Numerous studies have shown that combination of BAP and NAA has a stimulatory effect for multiple shoot elongation, although a decrease in shoot length on such media is reported (Nadgauda et al. 1983; Ganapathi et al. 1994; Babu et al. 1997). In contrast, the growth of cardamom cultures was impeded by NAA, resulting in shoots displaying restricted growth and browning after just 2 months of being cultured. Cultures initiated on media containing varied concentrations of BAP and 2.68 μM NAA exhibited favorable growth responses, although a slight browning was noticed at the basal part within the first month of cultivation. Callus formation was evident on the base of explants cultured on medium enriched with different concentrations of BAP and NAA levels surpassing 8.06 mM. The explants cultured on media containing higher NAA concentrations (8.06–10.75 μM) completely dried up following browning of the base after 2 months of culture (Malhotra et al. 2021).

4.3.3 Induction of In Vitro Rooting

Microshoots must be rooted in order to be established in soil. Prior research has frequently discovered that, in a hormone-free MS media, the plantlet's intrinsic auxin levels are sufficient to promote root growth (Reghunath and Bajaj 1992; Bajaj et al. 1993; Babu et al. 1997; Tyagi et al. 2009). All of the cytokinin and auxin media combinations that were evaluated showed rhizogenesis, and both full-strength and half-strength MS basal medium was shown to induce a satisfactory in vitro root induction response. On a medium supplemented with 17.14 μM IAA (indole-3-acetic acid), very few roots were detected, and all the plants cultivated on this medium dried up after 4 weeks. Cardamom plants showed a rooting frequency of 91.7% on half-strength MS medium, with roots appearing 10 days after culturing. On full-strength MS basal medium, the greatest number and longest roots—3.50 and 4.33 cm, respectively—were found (Malhotra et al. 2021).

4.3.4 Cell Suspension Cultures

Rhizomes of cardamom can be cultured in Murashige and Skoog (MS) medium supplemented with different combinations of plant growth regulators for callus induction. Under both dark and 12-h photoperiod circumstances, MS medium supplemented with different combinations of cytokinins, 6-benzylaminopurine (BAP) and kinetin (Kn), coupled with auxin, 2,4-dichlorophenoxyacetic acid (2,4-D), was maintained at room temperature at 25 ± 2 °C. After four subcultures that result in a fine cell suspension, the friable callus material that is created can be further subcultured in liquid media with the same PGR concentration to produce a cardamom cell suspension culture, which can then be used for downstream research. For a 12-h photoperiod, cultures must be kept in

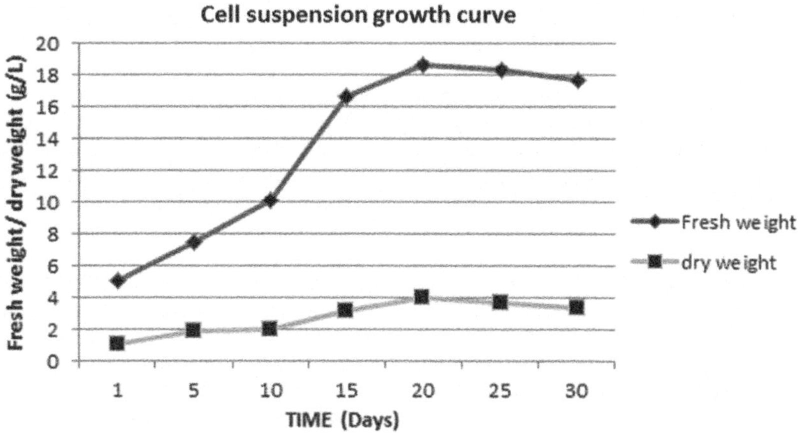

Fig. 4.2 Cardamom cell suspension growth curve depicting the fresh weight and dry weight of cells

250 mL flasks at 25 ± 2 °C and continuously shaken at 125 rpm. Every 20 days, 0.5 g of cell culture was subcultured into 60 ml of fresh nutritional medium. The maximal friable callus from the aforementioned hormone combination may then be subcultured in the same medium without Phytagel to yield a cell suspension with an active rate of proliferation (Dharan and Sabu 2023). The fresh weight and dried weight of the cells peaked at day 20, and subsequently either remained constant or decreased, according to an analysis of the growth pattern of the cells during a thirty-day period with five-day intervals (Fig. 4.2).

4.4 Conclusion

Micropropagation methods in small cardamom have been reported by many workers (Rao et al. 1982, 2000; Nadgauda et al. 1983; Kumar et al. 1985; Vatsya et al. 1987; Reghunath 1989; Reghunath and Bajaj 1992; Priyadarshan et al. 1992; Lukose et al. 1993; Nirmal Babu et al. 1997; Madhusoodanan et al. 2002; Malhotra et al. 2021). The approaches were extensive either through callusing, adventitious bud formation, proliferation of inflorescence primordium, immature inflorescence segments, immature capsules, and enhanced axillary branching. It has also been observed that cardamom buds can be encapsulated to produce synthetic seeds, which can be used as an alternate propagation method (Ganapathi et al. 1994; Sajina et al. 1997). Similarly, cardamom has been observed to benefit from different tissue culture techniques, including somatic embryogenesis (SE) and protoplast culture. Geetha et al. (2000) reported the successful isolation of protoplasts from cell suspensions and leaf tissues in the cardamom to microcalli stage. SE from rhizome callus followed by successful plantlet regeneration was reported by Manohari et al. (2008). However, more attention has to be given to raise micropropagated plantlets of cardamom with no variability to ensure uniform yield characteristics.

References

Babu N, Ravindran PN, Peter KV (1997) Protocols of micropropagation of spices and aromatic crops. Indian Institute of Spices Research Calicut, Kerala, India

Bajaj YPS, Reghunath BBR, Gopalakrishnan PK (1993) *Elettaria cardamomum* Maton (cardamom): aromatic compounds, in vitro culture studies, and clonal propagation. In: Bajaj YPS (ed) Biotechnology in agriculture and forestry, vol. 21 medicinal and aromatic plants IV. Springer-Verlag, Berlin Heidelberg, pp 132–147

Chempakam B, Sindhu S (2008) Small cardamom. In: Partasarathy V, Chempakam B, Zachriah TJ (eds) Chemistry of spices. Centre of Agriculture and BioScience International, Wallingford, UK, pp 41–58

Debergh PC, Zimmerman RH (1991) Micro-propagation technology and application. Kluwer Acedemic Publishers, Dordrecht

Dharan SS, Sabu KK (2023) Expression profiling of stress responsive genes in cell suspension of *Elettaria cardamomum* (L.) Maton under abiotic stress. J Plant Crops 51(1):23–30. https://doi.org/10.25081/jpc.2023.v51.i1.8188

Ganapathi TR, Bapat VA, Rao PS (1994) *In vitro* development of encapsulated shoot tips of cardamom. Biotechnol Tech 8:239–244

Geetha SP, Nirmal BK, Rema J, Ravindran PN, Peter KV (2000) Isolation of protoplasts from cardamom (*Elettaria cardamomum* Maton.) and ginger (*Zingiber officinale* Rosc.). J Spices Aromat Crops 9:23–30

George EF, Hall MA, Klerk GJ (2008) Plant propagation by tissue culture, 3rd edn, vol 1. Springer Publications, New York

Kumar KB, Prakashkumar P, Balachandran SM, Iyer RD (1985) Development of clonal plantlets from immature panicles of cardamom. J Plant Crop 13:31–34

Lukose R, Saji KV, Venugopal MN, Korikanthimath VS (1993) Comparative field performance of micropropagated plants of cardamom (*Elettaria cardamomum*). Indian J Agric Sci 63:417–418

Madhusoodanan KJ, Pradeep KK, Ravindran PN (2002) Botany, crop improvement and biotechnology of cardamom. In: Ravindran PN, Madhusoodanan KJ (eds) Cardamom—the genus *Elettaria*. Taylor & Francis Inc., London, pp 11–68

Malhotra EV, Kamalapriya M, Bansal S, Meena DP, Agrawal A (2021) Improved protocol for micropropagation of genetically uniform plants of commercially important cardamom (*Elettaria cardamomum* Maton). In Vitro Cell Dev Biol Plant 57:409–417

Manohari C, Backiarani S, Jebasingh T, Somanath A, Usha R (2008) Efficient plant regeneration in small cardamom (*Elettaria cardamomum* Maton), through somatic embryogenesis. Indian J Biotechnol 7:407–409

Nadgauda RS, Mascareenhas AF, Madhusoodanan KJ (1983) Clonal multiplication of cardamom (*Elettaria cardamomum* Maton) by tissue culture. J Plant Crop 11(1):60–66

Nirmal Babu K, Ravindran PN, Peter KV (1997) Protocols for micropropagation of spices and aromatic crops. Indian Institute of Spices Research, Calicut, India

Priyadarshan PM, Zachariah PK (1986) Studies on *in vitro* culture of cardamom (*Elettaria cardamomum* Maton; Zingiberaceae)—progress and limitations. In: Somers DA, Gengenbach BG, Biesboer DD, Hackett WP, Green CE (eds) Proceedings of VI international conference on plant tissue and cell culture, Minneapolis, Minnesota, USA, p 107

Priyadarshan PM, Kuruvilla KM, Madhusodanan KJ, Naidu R (1992) Effect of various media protocols and genotype specificity on micropropagation of cardamom (*Elettaria cardamomum* Maton.). In: Subba Rao NS, Rajagopalan C, Ramakrishna SV (eds) New trends in biotechnology. Oxford and IBH Publications, New Delhi, India, pp 109–117

Purseglove JW (1975) Tropical crops: monocotyledons, vol 1. Longman, Essex, p 530

Purseglove JW, Brown EG, Green CL, Robbins SRJ (1981) Spices, vol 2. Longman Inc., New York, p 813

Rao NS, Narayanaswamy S, Chacko EK, Swamy RD (1982) Regeneration of plantlets from callus of *Elettaria cardamomum* Maton. Proc Indian Acad Sci (Plant Sci) 91:37–41

Rao YS, Mary MK, Pradip KK, Madhusoodanan KJ, Potty SN, George CK (2000) Practical manual for micropropagation of some selected spices. Published by Indian Cardamom Research Institute, Spices Board, Cochin, 60 pp

Ravindran PN, Babu N (2002) Production of haploids of cardamom (*Elettaria cardamomum* Maton) through anther culture/microspore culture. Project report. Indian Institute Spices Research, Calicut

Reghunath BR (1989) *In vitro* studies on the propagation of cardamom (*Elettaria cardamomum* Maton). Ph.D. thesis, Kerala Agricultural University, Thrissur, India

Reghunath BR, Bajaj YPS (1992) Micropropagation of cardamom (*Elettaria cardamomum* Maton). In: Bajaj YPS (ed) High-tech and micropropagation III. Springer, Berlin, Heidelberg, pp 175–198

Reghunath BR, Gopalakrishnan PK (1991) Successful exploitation of *in vitro* culture techniques for rapid clonal propagation and crop improvement in cardamom. In: Proceedings of the Kerala science congress, Kozhikode, Kerala

Sajina A, Minoo D, Geetha P, Samsudeen K, Rema J, Nirmal Babu K, Ravindran PN, Peter KV (1997) Production of synthetic seeds in few spice crops. In. Edison S, Ramana KV, Sasikumar B, Nirmal Babu K, Eapen SJ (eds) Biotechnology of spices, medicinal and aromatic plants. Indian Society for Spices, Calicut, India, pp 65–69

Spices Board (1996) Annual report. ICRI, Spices Board, Kochi, India

Spices Board (2023) Annual Report 2022–23. Spices Board India, Ministry of Commerce and Industry, Government of India, Kochi, p 72

Sudharsan MR, Bhatt SS, Narayanaswamy M (1997) Variability in the tissue cultured cardamom plants. In: Edison S, Ramana KV, Sasikumar B, Nirmal Babu K, Eapen SJ (eds) Biotechnology of spices, medicinal and aromatic plants. Indian Society for Spices, Calicut, pp 98–101

Tyagi RK, Goswami R, Sanayaima R, Singh R, Tandon R, Agrawal A (2009) Micropropagation and slow growth conservation of cardamom (*Elettaria cardamomum* Maton). In Vitro Cell Dev Biol Plant 45:721–729

Vatsya B, Dinesh K, Kundapurkar AR, Bhaskaran S (1987) Large scale plant formation of cardamom (*Elettaria cardamomum*) by shootbud cultures. Plant Physiol Biochem 14:14–19

Genomics of Biotic Stress Tolerance in Small Cardamom

5

Sangeetha Elangovan, Julius Jeyaprakasam, and Jebasingh Tennyson

Abstract

Cardamom is an important spice crop, and its productivity is being highly influenced by various biotic stresses including viruses, bacteria, fungi, phytoplasma, nematodes and insects, some of which cause significant yield loss. The complete genetic information for cardamom is lacking but the availability of genetic information for a few cardamom-infecting pathogens, the differentially expressed mRNAs and miRNAs of cardamom with respect to various pathogens has improved our knowledge towards understanding the pathogen-plant interactions and facilitates the crop improvement for biotic stress resistance. A continuous genomics study against cardamom-infecting pathogens and otherorganisms is necessary to identify new genes that can provide resistance against pathogens. This chapter presents the genetic diversity of cardamom, the genotyping of cardamom pathogens and further deals with the identification of various genes through transcriptomic, proteomic and metabolomic means for the development of biotic stress-resistant cardamom.

Keywords

Cardamom · Pathogen · Biotic-stress · Genetic diversity · Molecular markers · Transcriptome · miRNA

5.1 Introduction

Generally, plants persistently meet a variety of environmental stresses that affect agricultural efficiency. Stress in plants refers to peripheral conditions that negatively have an effect on the growth, development or productivity of plants (Verma et al. 2013). Due to stress, a broad range of plant responses emerged like altered gene expression, cellular metabolism, changes in growth rates, crop yields, *etc*. Two primary categories of stress conditions occur in plants namely abiotic and biotic stress. Biotic stresses are of major concern because most of the pathogens are poorly understood and variability in plant physiology makes it difficult to develop control measures (Ayres 1992). Plants have a well-definite, fine-tuned defence system that responds to environmental pressure and pathogen assail as demanded by their sessile nature. Compared to

S. Elangovan · J. Jeyaprakasam · J. Tennyson (✉)
Department of Plant Sciences, School of Biological Sciences, Madurai Kamaraj University, Madurai, Tamil Nadu 625021, India
e-mail: jebasingh.biological@mkuniversity.ac.in

S. Elangovan
Department of Biotechnology, Thiagarajar College, Madurai, Tamil Nadu, India

K. K. Sabu and C. Kole (eds.), *The Cardamom Genome*, Compendium of Plant Genomes,
https://doi.org/10.1007/978-3-031-71829-8_5

animals, plants do not have portable defenders or adaptive immune systems to defend the pathogens but they depend only on the innate immunity of each cell and systemic signals emerging from the site of infection (Ausubel 2005). The primary mode of defence includes the physical barriers like rigid cell wall, cell lignification, waxes, epidermal cuticles and the chemical responses like ROS production, nitric oxide production and deposition of callose in the plasmodesmata which correlate with the size exclusion limit of the plasmodesmata. The main types of responses include pathogen-associated molecular pattern (PAMP)- triggered immunity (PTI) and effector-triggered immunity (ETI). These responses are the result of plant–pathogen interactions to develop resistance or suppression.

Small cardamom (*Elettaria cardamomum*) is traditionally known as the 'Queen of spices' which is the third most expensive spice in the world with a worldwide turnover of more than 10 billion USD (Khan et al. 2020). It is a perennial, herbaceous, rhizomatous, monocot plant that belongs to the family *Zingiberaceae* under the Zingiberales order. Cardamom is cultivated in tropical regions like India, Guatemala, Sri Lanka, Indonesia, Nepal, *etc*. India, the 'land of spices' once occupied a premier position in the global cardamom market, but now Guatemala is the largest producer of cardamom with an annual yield of 23,000 metric tons (Sasikumar et al. 2012). The widespread incidence of diseases adversely affected the cardamom industry and resulted in a substantial reduction in total production and productivity. Cardamom as a result of shade tolerance, promotes vegetative growth at 40–50% (Kumar et al. 2015). During the 10–15 years of its lifespan, the cardamom plant is exposed to various biotic stress factors such as viruses, bacteria, phytoplasia, fungi, nematodes and pests. The fungal diseases like capsule rot, rhizome rot, leaf blight, viral diseases like cardamom mosaic disease, cardamom vein clearing disease, chlorotic streak disease, nematode diseases like root knot, root grub and bacterial wilt disease are of major concern (Nair 2020).

Cardamom is an old spice crop that originated in the biodiversity-rich evergreen forests of the Western Ghats, India. Due to perennial nature and the continued cultivation of cardamom for centuries has resulted in considerable genetic diversity. The domestication of wild cardamom leads to the production of various cultivars significantly increasing the genetic diversity (Kuriakose et al. 2009). The present chapter accounts for the information about the genetic diversity of cardamom, genome-wide identification of molecular markers, cardamom-infecting organisms, genotyping of cardamom pathogens and genomic, proteomic and metabolomic approaches for the development of biotic stress resistant cardamom.

5.2 Genetic Diversity of Cardamom

Cardamom is indigenous to south India and Sri Lanka and has originated in the moist evergreen rainforests of the Western Ghats of south India (Kumar et al. 1997). Originally, cardamom pods were harvested from the cardamom growing in the forests and cultivation of cardamom began during 1803, and then selection of cardamom with respect to better agronomic traits began. Further, the cultivation was usually practiced in forests after clearing all the undergrowth where cardamom grows naturally. These practice gradually eliminated wild cardamom. Frequent cultivation of cardamom and targeted selection of peculiar morpho/ecotypes have resulted in the development of new varieties with genetic variations (Anjali et al. 2016). Basically, two varieties of cardamom plants are identified and they are *Elettaria cardamomum* Maton, a major variety comprised of wild indigenous types of Sri Lanka and *Elettaria cardamomum* Maton, a minor variety comprised of cultivars like *Mysore*, *Malabar* and *Vazhukka*. These types are grown in different tracts and are mostly identified based on vegetative and morphological characteristics. Cardamom varieties are highly location specific.

The genetic diversity or biodiversity relies on the phylogenetic variation in plants; the increase in the number of species and also the species turnover. The occurrence of diversity among different cultivars of cardamom is reported and most of the variability is intended for economic importance (Padmini et al. 2001). The large amount of variability in small cardamom germplasm occurs only in India (Prasath and Venugopal 2004). In Indian Cardamom Research Institute (ICRI), Kerala and its outstations, 814 cardamom accessions (Anisha et al. 2020) and in ICAR-Indian Institute of Spies Research (IISR) (NAGS), Appangala, Karnataka, more than 600 genotypes of cardamom have been maintained (Kumar et al. 2018) in which the morphological and potential yield cultivars were characterized for 117 accessions (Prasath et al. 2015). The high rate of diversity in cardamom might be the result of cross-pollination and seed propagation. The hybridization method is preferred over the selection method as it develops more genetic diversity, and helps to improve the cultivated cardamom (Babu et al. 2012).

5.3 Genome-Wide Identification of Molecular Markers

Genetic diversity analysis can be performed using morphological, biochemical and molecular markers. Among these, molecular markers offer consistent results despite the prevailing environmental circumstances. The identification of different molecular markers aids in the selection of a plant with desirable traits, known as marker-assisted selection (MAS). The MAS is an inevitable process in plant breeding. Molecular markers would be helpful to introduce new breeding strategies for recombination of different varieties and provide a genetic background for phenotypic variability (Cyriac et al. 2016). The screening of molecular markers like simple sequence repeats (SSR), inter-simple sequence repeats (ISSR), microsatellite markers, random amplified polymorphic DNA (RAPD), restriction fragment length polymorphism (RFLP), and single nucleotide polymorphism (SNP) helps to analyze the traits and to develop disease-tolerant varieties.

The amplification of microsatellite marker SSR from 20 genotypes of cardamom detected 22 different alleles. The SSR markers of class II type are abundant in cardamom with hexa- and hepta-nuclotide repeats. The flanking region of SSR is highly conserved and the allelic variation is due to the length variation in the repeats. The differentiation among the varieties from the same region is observed due to the different banding patterns in ECM14, ECM47a and ECMG28 loci (Cyriac et al. 2016). In addition, the SSR predicted from the expressed sequence tags (EST-SSR) of ginger and turmeric differentiated the wild cardamom genotype from the cultivator, and helps in the genetic diversity study and indicates the transferability of SSR markers among *Zingiberaceae* (Anjali et al. 2015; Sakthipriya and Sabu 2018). The SSR and SNP predicted from RNA sequence data would be helpful to analyze the molecular markers from the plants without a reference genome (Huang et al. 2016). The molecular profiling of small cardamom germplasm collections with 11 other species was performed based on the results obtained from RAPD, RFLP and ISSR markers (Babu et al. 2012). There are 220 polymorphic bands observed among the cultivars of cardamom in which varieties from the same origin show low genetic diversity (Babu et al. 2012). The low level of polymorphism among the genotypes is evident in the absence of diversity among the Indian population of cardamom (Cyriac et al. 2016). The ISSR profiling of 13 varieties of cardamom with 20 ISSR markers revealed the presence of 227 alleles, in which 197 were polymorphic. The distinct genotype was observed even between the varieties from the same geographic origin (Anisha et al. 2020) which is in contrast to the observations of earlier studies. Hence, in-depth studies on unexploited cardamom germplasm accessions are needed to develop a strategy for breeding programs (Anisha et al. 2020). The RNA sequence data from the control and CdMV-infected cardamom showed that 5317 SSR in differentially expressed genes of cardamom, 149,097 SNPs and 5120 Indels from the infected sample were observed and the higher numbers of SNPs are represented in plant immune responsive genes (Khan et al. 2020) easing the use of molecular markers to develop the disease resistant varieties.

5.4 Biotic Stress in Cardamom

Cardamom cultivation is affected by more than 25 diseases caused by fungi, bacteria, viruses, nematodes and insects (Fig. 5.1). These pathogens affect the plant in both nurseries and main plantations. On the basis of the severity of damage to cardamom, diseases are categorized into major and minor. Major diseases cause severe destruction with 50% yield loss (Nair 2020). The diseases caused in cardamom and the causative agents are listed in Tables 5.1, 5.2, 5.3 and 5.4. Nursery diseases include nursery leaf spot (caused by the fungus *Phyllosticta elettariae* and *Cercospora zingiberi*), nursery leaf rot (caused by fungi such as *Fusarium* and *Alternaria*) and damping off or seedling rot caused by *Pythium vexans* and *Rhizoctonia solani* and *Fusarium oxysporum.*

Fig. 5.1 Biotic stresses in small cardamom

Table 5.1 Viral diseases of small cardamom

Diseases	Virus/genus/family	Genome size	Transmission	References
Cardamom mosaic disease (*katte*)	*Cardamom mosaic virus* (CdMV)/*Maclura virus*/*Potyviridae*	~8.5 kb	*Pentalonia nigronervosa*	Jacob and Usha (2001) and Elangovan et al. (2019)
Chlorotic streak disease	*Banana bract mosaic virus* (BBrMV)/*Potyvirus*/*Potyviridae*	9.3 kb	*Pentalonia nigronervosa*	Siljo et al. (2012) and Bhat et al. (2018)
Cardamom necrosis disease	*Nilgiri necrosis virus*/*Carlavirus*/*Betaflexiviridae*	–	Through infected rhizomes	Sridhar et al. (1990)
Cardamom vein clearing disease (*kokke kandu*)	*Cardamom vein clearing virus* (CdVCV)/*Nucleorhabdovirus*/*Rhabdoviridae*	13.3 kb	*Pentalonia caladii*	Bhat et al. (2020)

Table 5.2 Bacterial and phytoplasma diseases of small cardamom

Disease	Causal organism	Genome size	References
Bacterial wilt	Race 4/biovar 3 of *Ralstonia solanacearum* CaRs_Mep	5.7 Mb	Kumar et al. (2017)
Shoot proliferation	*Candidatus Phytoplasma australasia*	1.25 kb	Mishra et al. (2019)

Table 5.3 Fungal diseases of small cardamom

Diseases	Causal organism	References
Azhukal or capsule rot	*Phytophthora meadii*	Menon et al. (1972)
Damping off or seedling rot	*Pythium vexans*	Thomas (2000)
Clump rot or rhizome rot	*Pythium vexans*, *Rhizoctonia solani* and *Fusarium oxysporum*	Mayne (1942)
Leaf blight	*Phytophthora meadii*, *Colletotrichum gloeosporioides*	Praveena et al. (2013)
Leaf blotch	*Phaeodactylum alpiniae*	Agnihothrudu (1968)
Leaf spots		
1. Sphaceloma *leaf spot*	1. *Sphaceloma cardamomi*	1. Muthappa (1965)
2. Cercospora *leaf spot*	2. *Cercospora zingiberi*	2. Rangaswami et al. (1968)
3. Glomerella *leaf spot*	3. *Glomerella cingulata*	3. Nair (1979)
Leaf rust	*Phakospora elettariae*	Thirumalachar (1943)
Anthracnose or brown spot	*Colletotrichum gloeosporioides*	Bhai et al. (1988)

Table 5.4 Nematode and insects/pests diseases of small cardamom

Diseases	Causal organism	References
Nematode disease	Root knot nematode: *Meloidogyne incognita* (plant-parasitic roundworm)	Ali (1982)
Insects/pests	1. Cardamom thrips: *Sciothrips cardamom* 2. Shoot/panicle/capsule borer: *Conogethes punctiferalis* (*Lepidoptera pyralidae*)	Thomas (2001)

5.4.1 Viral Diseases

Among the few viral diseases (Table 5.1), mosaic disease incited by *Cardamom mosaic virus* (CdMV) is considered a major production constraint encountered by the cardamom industry. The CdMV transmission is caused primarily by the aphid *Pentalonia nigronervosa* and 12 other aphid species (reviewed in Elangovan and Jebasingh 2023). Crop losses due to the mosaic disease depend on the growth stage at the time of infection. When the plants are infected at early stages of growth, the loss will be almost total, while late infection results in a gradual decline in productivity. Besides this dreadful disease, chlorotic streak caused by a strain of *Banana bract mosaic virus* (BBrMV) is an emerging disease of recent origin, which could possibly pose a threat to the cardamom industry in the future. Another viral disease, *kokke kandu* caused by *Cardamom vein clearing virus* has become a threat to cardamom cultivation in a few endemic pockets of Karnataka. Nilgiri necrosis disease is another disease that is prevalent in a few pockets in Nilgiris, Tamil Nadu. The intensity of this disease was found to be less than 1% in all the affected plantations (Venugopal 1995).

5.4.2 Bacterial Disease

Bacterial wilt disease in cardamom (Table 5.2) is caused by *Ralstonia solanacearum* which is known to infect all *Zingiberaceae* family members (Kumar et al. 2016). There are five races of *R. solanacearum* exist in the environment which infects more than 450 plant species. The occurrence of bacterial wilt disease in small cardamom was identified in 2011. *Ralstonia solanacearum* CaRs_Mep, a race 4/biovar 3 pathogen, was isolated from bacterial wilt affecting small cardamom plants originating from Meppadi, Wayanad district of the southern Indian state of Kerala. The symptoms of the disease include the rolling or curling of the leaves inward and the water-soaked dark brown lesions in the collar region (Kumar et al. 2012).

5.4.3 Phytoplasma Disease

In cardamom, the shoot proliferation disease is caused by the phytoplasma *Candidatus Phytoplasma australasia* (Table 5.2). Phytoplasma is a bacterium without cell wall, which multiplies in plant phloem and transmits through insects. There are very few phytoplasma diseases in spice crops reported all over the world. In the Njallani variety of cardamom, Idukki district, Kerala, excessive shoot proliferation, small panicles and short capsules were observed and later characterized to be phytoplasma disease. The disease incidence is found to be 20–30% and the vector for transmission is not yet known (Mishra et al. 2019).

5.4.4 Fungal Diseases

Major fungal diseases (Table 5.3) causing considerable crop loss are capsule rot, rhizome rot (clump rot), seedling rot (damping off), leaf blight, leaf blotch and nursery leaf spots. Capsule rot is locally known as Azhukal disease, caused by *Phytophthora meadii*. It is the most severe fungal disease of cardamom and a major constraint in the successful cultivation of cardamom. Symptoms of rotting are observed on the fruits or capsules. Rhizome rot disease is also known as clump rot caused by *Pythium vexans*, *Rhizoctonia solani* and *Fusarium oxysporum*. Its symptoms include pale-yellow colour in the foliage, wilting, premature death of older leaves and the collar portion of the aerial shoots becoming brittle. Leaf blight disease or Chenthal disease is caused by *Colletotrichum gloeosporioide*. Symptoms of this disease on the foliage as water-soaked rectangular lesions, which subsequently elongate to form parallel arranged streaks. Leaf blotch is caused *Phaeodactylium alpiniae* produces symptoms of water-soaked lesions in leaves. Various types of leaf spots are caused by a variety of fungal pathogens—*Sphaceloma cardamom*, *Cercospora zingiberi* and *Glomerella cingulata*. Leaf rust is caused by *Phakospora elettariae* which produces small pin-head white pustules on leaves that later become rusty brown in appearance. A sooty mold infection on leaves of cardamom occurred when the crops grew under the shade tree *Cedrella loona*.

5.4.5 Nematode Diseases

Heavy loss of a crop could be brought about by nematode infestation. Among 20 different nematodes (Table 5.4), the root knot nematode (*Meloidogyne incognita*) causes the most damage to cardamom. The root knot nematode is widely observed in all cardamom-growing regions, while the lesion nematode (*Pratylenchus coffeae*) and the burrowing nematode (*Radopholus similis*) are observed in mixed plantations. Nematodes cause aerial damages such as stunting, reduced tillering, resetting and narrowing of leaves, thickening of veins yellow banding of leaf blades, drying of leaf tips or leaf margins, dropping of immature fruits and underground symptoms like root galling and spherical-ovoid swellings in tender roots. Severe infestation can result in crop losses of up to 80%.

5.4.6 Insect Diseases

Small cardamom is infested by several insects (Table 5.4). Among the insect pests, thrips, capsule/panicle/shoot borer and root grubs are the major players. Minor pests include white fly, shoot fly, scale insects, early capsule borer, hairy caterpillar, lacewing bug, red spider mites, etc. Thrips (*Sciothrips cardamomi* (Ramk.)) is the most destructive and persistent pest of cardamom found in all cardamom-growing areas. Shoot/Panicle/Capsule borer [*Conogethes punctiferalis* (Guen.)] causes infestation of borer on capsule, panicle and shoot is a serious problem on cardamom. Early capsule borer (*Jamides alecto*) is a minor pest of cardamom. Root grub [*Basilepta fulvicorne* (Jacoby)] is a serious pest that damages the roots of cardamom.

5.5 Multilocus Sequence Typing of Cardamom-Infecting Bacterial Genome

Multilocus sequence typing (MLST) is a molecular technique for the typing of multiple loci that helps to characterize the bacterial isolates based on the allelic variation in housekeeping gene sequence among the isolates (Maiden et al. 1998). The allelic variation or relatedness observed through sequencing of six or seven loci or internal fragments of housekeeping genes helps to assign the sequence type or lineage by comparing with other sequences (Dingle and MacCannell 2015). Bacterial wilt disease in small cardamom is caused by *Ralstonia solanacearum*. There are five races of *R. solanacearum* that exist in the environment which infect more than 450 plant species (Kumar et al. 2005). The occurrence of bacterial wilt disease in small cardamom was identified on 2011. The *R. solanacearum* was isolated from the infected cardamom and the BLAST analysis of the rDNA sequence showed a 100% match with the complete genome of *R. solanacearum* of tomato isolate. 16S and 23S rDNA and *recN* gene sequences of the cardamom strain showed 99% identity with the tomato isolate. Further, the BOX-PCR was performed to compare the cardamom strain (CaRs-Mep) with the strains obtained from wilt disease-infected ginger (GRs-Tms), paprika (CRs-Per) and potato (PRs-Pun). Based on the similarity matrix obtained from Box-PCR analysis revealed that the CaRs-Mep is closely related to GRs-Tms than to CRs-Per and PRs-Pun. The genotyping using Multilocus Sequence Typing (MLST) of cardamom and other three strains for eight alleles (*ppsA*, *gapA*, *gdhA*, *adk*, *gyrB*, *fliC*, *hrpB* and *egl*) showed 100% identity among cardamom and ginger strains. Based on these results, the cardamom strain is characterized to being similar to the strain responsible for the wilt disease in ginger (Kumar et al. 2012). *Zingiberaceae* family members are infected with race 4/biovar 3 or biovar 4 of *R. solanacearum* in which biovar 3 is predominant in India. The phylogenetic analysis of 33 isolates of *R. solanacearum* indicates that the cardamom strain belongs to the biovar 3 and phylotype 1 (Kumar et al. 2013).

Elettaria cardamomum shoot proliferation (ECSP) disease is caused by the phytoplasma *Candidatus Phytoplasma australasia*, the only known phytoplasma disease in cardamom. The observed symptoms of ECSP include excessive shoot proliferation with short panicles and degenerated capsules. The ECSP was identified in Njallani variety, Nedumkandam Panchayat, Idukki, Kerala, in 2017. The disease incidence was found to be 20–30%. The 16s rRNA sequencing confirmed that the ECSP strain showed 100% identity with the GenBank submitted Ca. *P. australasia* strains. The MLST-based gene sequence homology analysis was performed to compare the ECSP strain with the GenBank submitted sequences. The four gene sequences of ECSP—*secA*, *secY*, *tuf* and *SAP11* shared 99.37%, 99.6–100%, 98–100% and 99.44% identity with other strains, respectively. The phylogenetic tree and the RFLP analysis confirmed the ECSP strain belongs to the 16SrII-D subgroup of phytoplasma (Mishra et al. 2019).

5.6 Strategies Adopted to Develop Cardamom Against Various Pathogens

5.6.1 Micropropagation and Disease Resistant

In cardamom, the in vitro methods of clonal propagation (Babu et al. 1997), micropropagation (Nadgauda et al. 1983), regeneration of plantlets from callus (Rao et al. 1982), somatic embryogenesis (Manohari et al. 2008) are developed successfully. Geetha et al. (2000) isolated and cultured the cardamom protoplasts from leaf mesophyll tissue as cell suspension culture. Micropropagation is helpful to generate virus-free plants. The screening of cardamom mosaic disease resistance in micropropagated cardamom identified 67 virus-free soma clones and few of the soma clones were tolerant to the cardamom mosaic disease (Peter et al. 2001). The cardamom culturing methods would help to develop virus-free plants and in some cases disease-tolerant varieties (Table 5.5), however, the limited rate of success and time consumption make these methods unsuitable.

5.6.2 Genetic Transformation in Cardamom

The introduction of the complete genome or protein of the pathogen in plants through transformation methods will be helpful to study the immune responses elicited by plants and to know the host protein interacting with the pathogen. The *GUS* gene transformation using the biolistics method was attempted and the transient expression of *GUS* gene was observed in embryogenic callus (Babu et al. 1998). The efforts to develop the CdMV-resistant cardamom by transforming CdMV coat protein (CP) with and without 3′UTR (Backiyarani et al. 2007) and Nuclear inclusion protein b (NIb) (Jebasingh et al. 2017) using the biolistics method of transformation would facilitate the development of pathogen-derived resistance in cardamom. Interestingly, the screening of NIb transgenic cardamom by southern hybridization using NIb probe detected the integration of a fragment of NIb (182 bp corresponding to the 3′ sequence of NIb) in control cardamom (Jebasingh et al. 2017). The viral gene integration in the plant genome may be attributed to the natural evolution of resistant varieties.

5.6.3 Genomic Approaches in the Development of Biotic Stress Resistant Cardamom

The limited success rate in resistant variety development in conventional methods necessitates the intervention of genomic approaches in order to manage the prevalence of diseases in small cardamom. The discovery of resistance genes and the elucidation of the mechanisms

Table 5.5 Disease-resistant/tolerant varieties of cardamom

Diseases	Resistant/tolerant variety/germplasm accessions	Detection method	Source/References
Rhizome rot	• IISR Avinash (tolerant)	• Selection from open pollinated (OP) seedling of Malabar type	• IISR, Calicut Venugopal et al. (2006)
	• IISR Suvasini (tolerant)	• Selection from OP progeny of Malabar type	• IISR, Calicut
	• ICRI 3 (tolerant)	• Selection from Malabar type	• ICRI, Spices Board, Sakleshpur
	• TKD 4	• Clonal selection from Malabar type	• ICRI, Spices Board, Thadiyankudisai
	• Highly resistant FGB 1, FGB 3, FGB 8, FGB 9, FGB 13, FGB 21, FGB 22, FGB 26, FGB 28, FGB 29, FGB 30, FGB 45, FGB 52, FGB 60, FGB 118 Resistant FGB 5, FGB 19, FGB 27, FGB 31, FGB 33, FGB 34, FGB 49, FGB 50, FGB 58, FGB 107, FGB 117, FGB 119	• Field screening of the cardamom germplasm accessions by calculating the percent disease index (PDI) from IISR, Appangala	• Sharon et al. (2020)
	• Highly resistant IC 547190, IC 547198, IC 547199, IC 547200, IC 547201, IC 547215, IC 584093, IC 584075, IC 584072, IC 584069, IC 584068, IC 584067, IC 584087, IC 547224, IC 584081, IC 547226, IC 547186, IC 584084, IC 547183, IC 547210, IC 547182, IC 547195, IC 547196, IC 547197, IC 547202, IC 547203, IC 547204, IC 547207, IC 547208 Resistant IC 584070, IC 584083, IC 547208, IC 547188, IC 547185	• Screening of different genotypes from IISR, Appangala by calculating the PDI	• Kumar et al. (2018)
	• IC 547222, IC 547223, IC 349645, IC 349649, IC 547158, IC 349637	• Screening of cardamom accessions, IISR, Appangala	• Praveena et al. (2013)
Capsule rot	• ICRI 2, ICRI 5, ICRI 6 (tolerant)	• Clonal selection from germplasm collection, Mysore type	• ICRI, Spices Board, Myladumpara
	• Thiruthali (moderately tolerant)	• Natural hybridization among Njallani, Malabar and Vazhukka type	• Improved variety developed by T. P. Joseph, a farmer

(continued)

Table 5.5 (continued)

Diseases	Resistant/tolerant variety/germplasm accessions	Detection method	Source/References
Leaf blight	• Highly resistant FGB 1, FGB 2, FGB 3, FGB 4, FGB 5, FGB 7, FGB 8, FGB 9, FGB 11, FGB 13, FGB 14, FGB 15, FGB 18, FGB 19, FGB 21, FGB 22, FGB 24, FGB 25, FGB 27, FGB 28, FGB 30, FGB 31, FGB 37, FGB 39, FGB 44, FGB 46, FGB 52, FGB 53, FGB 55, FGB 56, FGB 58, FGB 60, FGB 67, FGB 87, FGB 113 Moderately resistant FGB 6, FGB 10, FGB 12, FGB 16, FGB 17, FGB 20, FGB 23, FGB 26, FGB 29, FGB 32, FGB 33, FGB 34, FGB 35, FGB 36, FGB 41, FGB 45, FGB 47, FGB 48, FGB 50, FGB 54, FGB 57, FGB 59, FGB 63, FGB 64, FGB 65, FGB 66, FGB 68, FGB 71, FGB 72, FGB 73, FGB 75, FGB 76, FGB 78, FGB 84, FGB 86, FGB 89, FGB 90, FGB 94, FGB 97, FGB 98, FGB 118, FGB 119, FGB 120	• Field screening of the cardamom germplasm accessions by calculating the PDI from IISR, Appangala	• Sharon et al. (2020)
	• Highly resistant IC 547189, IC 584073, IC 584083, IC 584086, IC 584087, IC 584088, IC 547224, IC 584081, IC 547185, IC 547225, IC 547226, IC 547207, IC 547208, IC 547188, IC 547187, IC 584084, IC 547183, IC 547202, IC 547203, IC 547204, IC 547210, IC 547213 Moderately resistant IC 547190, IC 547199, IC 547200, IC 547201, IC 547214, IC 547215, IC 547162, IC 547169, IC 584096, IC 584095, IC 584094, IC 584091, IC 584090, IC 584089, IC 584079, IC 584072, IC 584069, IC 584068, IC 547184, IC 584085, IC 584080, IC 584082, IC 547182, IC 547206, IC 547212, IC 547207	• Screening of different genotypes from IISR, Appangala by calculating the PDI	• Kumar et al. (2018)
	• Highly resistant IC 349613, IC 349588 Resistant IC 349368, IC 349402, IC 349411, IC 349484, IC 349646, IC 349410, IC 349440, IC 349453 Moderately resistant IC 349619, IC 349333, IC 349344, IC 349353, IC 349401, IC 349460, IC 349532, IC 349539, IC 349552, IC 349582, IC 349639, IC 349472, IC 349592, IC 349370, IC 349421, IC 349437, IC 349469, IC 349517, IC 349528, IC 349534, IC 349524	• Screening of cardamom accessions, IISR, Appangala	• Praveena et al. (2013)

(continued)

Table 5.5 (continued)

Diseases	Resistant/tolerant variety/germplasm accessions	Detection method	Source/References
Shoot and capsule borer	• Mudigere 1 (tolerant)	• Clonal selection from Malabar type	• Regional research station, Mudigere
	• Appangala-2 (tolerant)	• Hybridization of Appangala 1 and NKE-19	• IISR, Kerala Kumar et al. (2017)
	• IISR-Suvasini, IISR-Avinash (tolerant)	• Selection from OP seedling of Malabar type	• IISR, Calicut
	• IISR-Vijetha (tolerant)	• Clonal selection from field katte resistant plant (NKE-12), a Malabar type	• Venugopal (2002)
	• Mudigere 1 (moderately tolerant)	• Clonal selection from Malabar type	• Regional research station, Mudigere
	• PV 1 (tolerant)	• Selection from Walayar collection, a Malabar type	• Cardamom research station, Pampadumpara
	• ICRI 6 (moderately tolerant)	• Clonal selection from germplasm collection, Mysore type	• ICRI, Spices Board, Myladumpara
Stem borers and root grubs	• PV 2 (tolerant)	• Selection from OP seedlings of PV 1, a Vazhukka type	• Cardamom research station, Pampadumpara Madhusoodanan (2012) and Josephrajkumar et al. (2002)
	• Thiruthali (resistant)	• Natural hybridization among Njallani, Malabar and Vazhukka type	• Improved variety developed by T. P. Joseph, a farmer
Cardamom mosaic disease	• IISR Vijetha (resistant)	• Clonal selection from field katte resistant plant (NKE-12), a Malabar type	• Venugopal (2002)
	• IISR-Avinash (moderately resistant)	• Selection from open pollinated (OP) seedling of Malabar type	• IISR, Calicut Venugopal et al. (2006)
	• Appangala-2 (resistant)	• Hybridization of Appangala 1 and NKE-19	• IISR, Kerala Kumar et al. (2017)
	• NKE 3, NKE 4, NKE 5, NKE 8, NKE 9, NKE 11, NKE 12, NKE 19, NKE 26, NKE 27, NKE 28, NKE 31, NKE 32, NKE 34, NKE 71, NKE 72, NKE 78	• Field screening of natural katte escape varieties	• Venugopal (1999)

(continued)

Table 5.5 (continued)

Diseases	Resistant/tolerant variety/germplasm accessions	Detection method	Source/References
Thrips	• IISR-Vijetha (tolerant)	• Clonal selection from field katte resistant plant (NKE-12), a Malabar type	• Venugopal (2002) and Madhusoodanan (2012)
	• IISR-Suvasini (tolerant)	• Selection from OP progeny of Malabar type	• IISR, Calicut
	• Mudigere-1, Mudigere-3 (tolerant)	• Clonal selection from Malabar type	• Regional research station, Mudigere
	• ICRI 6 (tolerant)	• Clonal selection from germplasm collection, Mysore type	• ICRI, Spices Board, Myladumpara
	• PV 1 (tolerant)	• Selection from Walayar collection, a Malabar type	• Cardamom research station, Pampadumpara
	• PV 2 (tolerant)	• Selection from OP seedlings of PV 1, a Vazhukka type	• Cardamom research station, Pampadumpara
Hairy caterpillars, white grubs	• Mudigere 2	• Clonal selection from OP Malabar type	• Regional research station, Mudigere

involved improve the plants resistance to pathogens (Babu et al. 2013). The structural, functional and comparative genomics help in the discovery of the disease resistant genes and their sequence, allelic variation, functional role, expression level at a specific period of time or under stress conditions, *etc* (Michelmore 2000).

5.6.4 Genome Sequencing Approach to Develop Resistance Varieties

Based on the transcriptome of cardamom and genome of cardamom-infecting pathogens and the transcriptome changes corresponding to the pathogens would pave the way to develop disease-tolerant varieties. The cardamom has a diploid genome, based on the 2C nuclear DNA content derived from flow cytometry, Anjali et al. (2016) calculated the genome size of the small cardamom which is found to be 2806 Mbp. But the complete chromosomal genome sequence of cardamom is not available yet, and only limited genomic information is available for a few cardamom pathogens. The complete genome sequence is available for *R. solanacearum* (CaRs-Mep) and for the viruses (BBrMV, CdMV) and *Cardamom vein clearing virus* (CdVCV). The complete genome sequence for cardamom-infecting pathogens is either predicted with the traditional Sanger sequencing or with advanced Next Generation Sequencing (NGS) methods. The complete genome sequence of *R. solanacearum* (CaRs-Mep) is determined Illumina NGS sequence has a 5.7 Mb size genome with 5127 genes includes functionally characterized 2985 genes in biological processes, 2813 genes in molecular functions and 1811 genes in cellular components (Kumar et al. 2017).

The complete genome sequence of cardamom-infecting viruses namely BBrMV, CdMV and CdVCV is available till date. The BBrMV infection is first reported in bananas (Magnaye and Espino 1990) and later in flowering ginger (Wang et al. 2010) and cardamom (Siljo et al. 2012) and the spread of the disease is through *Pentalonia nigronervosa*. The complete genome sequence of the BBrMV-cardamom isolate was determined from the nine overlapping clones with Sanger sequencing. The BBrMV genome consists of 9372 nucleotides which encode a large polyprotein with 3124 amino acids. The polyprotein is cleaved by three virus-encoded proteases to release ten mature proteins: P1, Helper component protease (HC-Pro), P3, 7K, Cytoplasmic inclusion (CI), 9K, Viral protein genome-linked (VPg), Nuclear inclusion protein a (NIa) protease, Nuclear inclusion protein b (NIb), Coat protein (CP). An additional coding ORF, termed PIPO, overlaps the P3-encoding region of the polyprotein ORF in the +2 reading frame due to the RNA Polymerase slippage (Bhat et al. 2018).

The complete genome sequence of other cardamom-infecting viruses CdMV and CdVCV are determined with the NGS from the transcriptome of cardamom infected with the respective viruses. The first complete genome sequence of CdMV is detected from the transcriptome of CdMV-infected cardamom. The size of the CdMV genome is 8.2 kb which encodes a large polyprotein with 2636 amino acids. The polyprotein is processed into nine individual proteins—HC-Pro, P3, 7K, CI, 9K, VPg, NIa, NIb and CP with the help of two virus-encoded proteases—HC-Pro and NIa protease. The other small ORF, P3N-PIPO also encodes from the genome of CdMV. The overlapping clones covering the entire genome of CdMV are constructed to characterize the genome and to develop an infectious CdMV clone (Elangovan et al. 2019). The CdVCV genome was detected from the aphid-inoculated plants showing typical cardamom vein clearing disease symptoms. The overlapping contigs produced in the small RNA and RNA sequencing assembled together to derive the 13.3 kb genome sequence of CdVCV. The genome consists of six ORF, which encode six proteins namely nucleocapsid (N), phosphoprotein (P), movement protein (P3), matrix protein (M), glycoprotein (G), and polymerase (L) (Bhat et al. 2020). The functional characterization of the genome sequence of the cardamom-infecting pathogens will assist in developing a resistant variety of cardamom against these pathogens.

5.6.5 Metagenome Analysis to Predict Novel Pathogens

The detection of microbial communities present in the environmental samples through sequencing is called metagenomics. In the case of plant viruses, the individual plant can be used as a metagenomic source (Roossinck et al. 2010). Plant viruses are characterized based on the symptoms that are developed in the plant. Some viruses develop latent symptoms, and in some mild strains of virus, the symptoms are often masked (Hildebrand 1958; Roossinck 2010). Virus detection from these samples is possible with metagenomics. Sidharthan et al. (2021) analyzed the two cardamom transcriptome datasets (Mathew et al. 2019; Khan et al. 2020) available in a public database and identified two novel viral sequences. The virus sequences similar to the members of *Potexvirus* and *Polerovirus* were identified and tentatively named Cardamom virus X and Cardamom polerovirus, respectively. The infection status of these two reported viruses has to be studied further.

5.6.6 Transcriptome Sequencing Approach to Study the Pathogen Responsive Genes

5.6.6.1 Cardamom Transcriptome Respect to Various Pathogens

The transcriptome sequencing compares the differentially expressed genes (DEGs) in control and infected plants. Khan et al. (2020) performed the transcriptome analysis of control vs CdMV-infected cardamom and recorded 5317 DEGs. The DEGs represent the genes and pathways involved in the virus-host interaction. The PAMP-associated receptor-like protein kinases (RLK) were found to be upregulated in CdMV-infected plants. The largest RLK family containing leucine-rich repeat (LRR-RLK) interacts with the pathogen protein through the LRR and affects the signalling response. In the CdMV-infected plant, the LRR receptor-like serine, threonine-kinase, receptor-like serine threonine-kinase SD1–8 were found to be upregulated. These upregulated genes can be used as candidate genes to develop a CdMV-resistant cardamom plant using genetic engineering or a transgene approach. The upregulation of transcription factors, namely bHL Hand MYB indicates the role of these transcription factors in the development of disease-resistant crops (Khan et al. 2020; Alves et al. 2014; Pireyre and Burow 2015).

Capsule rot is a fungal disease caused by *Phytophthora meadii*. The total RNA from leaf tissues of control and infected cardamom was used for the transcriptome sequencing. The raw data of control and infected have been deposited in the NCBI SRA database under the accessions SRX251235 and SRX2512358, respectively. Further comparison of the differential expression profile between the control and infected will be useful to identify the overexpressed R genes, which will be used to develop a resistance for *P. meadii* (Mathew et al. 2019).

5.6.6.2 Small Non-coding Sequencing with Respect to Various Pathogens

miRNAs are small non-coding RNAs with size ranges from 20 to 24 nucleotides (Jones-Rhoades et al. 2006) which is evolutionarily conserved across the different lineages in the plant kingdom. The miRNA is involved in post-transcriptional gene regulation by binding with the target mRNA which results in a degradation or translational suppression. The miRNAs have a significant role in plant development and responses during various biotic and abiotic stress conditions. The majority of miRNAs move from one tissue to another under stress conditions which have a role in stress response and resistance to pathogens (Sun 2012). The role of plant miRNA in defense against a pathogen and counter defense mechanism of pathogen miRNA clearly indicates the importance of miRNA in plant–pathogen interaction (Park et al. 2002; Pumplin and Voinnet 2013). Many agricultural crops utilize RNA interference or miRNA-mediated gene silencing mechanisms to defend themselves from pathogen attacks. The significant role of miRNA in plant defense responses to virus infection and virus counter defense has been used in crop improvement (Liu et al. 2017).

Small RNA sequencing using Ion Proton NGS technology has been used to predict the conserved and novel miRNAs of wild and cultivar cardamom which provides insights into the role of cardamom miRNAs under biotic and abiotic stress conditions (Nadiya et al. 2019; Anjali et al. 2019). In cardamom, under drought stress 17 differentially expressed miRNAs including drought stress target genes regulating miR156l and miR169c were identified (Anjali et al. 2019). miRNA expression in wild cardamom (TBGT86201) was compared with five cultivars with high yield and disease tolerance (TBGT86202, TBGT86203, TBGT86204, TBGT86205, TBGT86206). A total of 161 conserved miRNAs and totally 14 novel miRNAs (wild-9, cultivar-5) were identified from wild and cultivar cardamom. About 19 miRNAs are differentially expressed in cultivar cardamom compared with the wild in which 17 miRNAs were upregulated and 2 were downregulated (miR171b, miR529). In cultivars, greater than two-fold increase of certain miRNAs namely miR169, miR171 and miR166 were observed. The prediction of stress-responsive and defense response miR169, Eca-w01, Eca-w09 from wild and cultivar cardamom insists on the role of miRNA during biotic and abiotic stress conditions. The target gene prediction revealed that the targets of the miRNA include transcription factors, stress-responsive genes, defense response genes including NAC domain-containing proteins, LRR protein kinase family and bacterial resistance genes (Nadiya et al. 2019).

The virus also encodes miRNA which may have a remarkable role during the virus infection process. The biogenesis of miRNA takes place after the virus enters the host nucleus and transcribes into pri-miRNA (Pfeffer et al. 2004; Gao et al. 2012). The occurrence of miRNA in the sequence of BBrMV-Banana Indian isolate was predicted through computational approaches. In BBrMV, seven miRNAs were predicted and miRNA2 was experimentally validated from BBrMV infected banana (Sankaranarayanan et al. 2020). Since BBrMV-Banana isolate shows 96.7% identity with Indian BBrMV-Cardamom isolate (Bhat et al. 2018), the validation of BBrMV miRNA may provide better insights into cardamom crop improvement.

5.6.7 Proteome and Interactome

The understanding of the defense-related protein structure, functions and protein–protein interaction (PPI) or protein-nucleic acid interactions are prerequisite to study the nature of an infection. The PPI of plant and pathogen delineate the proteins involved in the defense system. The structural prediction of proteins of pathogen and host using computational approaches is widely used to study the interaction network. The structure of CdMV NIa (Jebasingh et al. 2013) and HC-Pro from BBrMV-Banana isolate (Anuradha et al. 2015) were predicted with homology modelling. The structure functionally characterizes the protein and enlightens to study the interaction mechanism. Insect cuticle protein sequence detected in the CdMV-infected cardamom transcriptome (Khan et al. 2020) helps to study the interaction that occurs in virus transmission. There is not much information about the protein structure and protein interaction study in cardamom pathogens. The interaction of CdMV VPg with the cardamom histone 3 and histone 4 proteins clearly indicates the CdMV protein accessibility towards the host genome (Palani et al. 2020). Further study about VPg-cardamom histone interaction will designate the role of VPg in defense response.

5.6.8 Role of Metabolites in Resistant Variety Development

Melatonin (N-acetyl-5-methoxytryptamine) is a multiuse signalling molecule that is widely present in living organisms. The biosynthesis of melatonin occurs in plants from the tryptophan pathway. In plants, an increase in melatonin concentration can be altered during biotic and abiotic stress conditions (Debnath et al. 2019). The availability of melatonin in *Elettaria cardamomum* seed is 15 nanograms per gram (Reiter and Tan 2002). Melatonin has antiviral activity which is also an anti-oxidant that has a protective role during stress conditions (Arnao and Hernández-Ruiz 2019). The exogenous application of melatonin in *Tobacco mosaic virus* (TMV)-infected *Nicotiana glutinosa* and *Solanum lycopersicum* activates the salicylic acid pathway and increases the expression of plant resistance genes PR1 and PR5 which reduces the concentration of TMV (Zhao et al. 2019). The application of melatonin in the shoot tip culture of *Malus domestica* infected with *Apple stem grooving virus* (ASGV) eradicates the virus (Chen et al. 2019). The antiviral activity of cardamom is not yet reported. The secondary metabolites have a major role in plant defense mechanisms. The transcriptome data from CdMV-infected cardamom showed the upregulation of Cytochrome P450 which controls the production of polyphenolic flavonoids which is known for its antiviral activity (Khan et al. 2020). The GC-MS study is also helpful to reveal the metabolites of cardamom. The chief constituents in the seeds of cardamom are α-terpinyl acetate, 1, 8-cineole and α-terpineol are responsible for aroma (Asghar et al. 2017; Alam et al. 2021). Upon pathogen attack, the quality and aroma of cardamom capsules are affected. The detection of involvement of pathogens in the biosynthesis of these metabolites may be helpful to develop disease-resistant varieties.

5.7 Future Perspective

Cardamom is the third most expensive spice, affected by various biotic and abiotic factors, reducing the yield. Specifically, the disease management of cardamom is a major challenge due to limited success in resistant variety development thus, an alternative genomic approach is needed to supplement such efforts. The application of genomics on cardamom plants for biotic stress resistance is at infant level but serious application of genomics in cardamom is needed for protecting the crop from the pathogen to improve productivity. Few transcriptomic reports have been available for this crop with response to abiotic (drought) and biotic (virus) stresses. By RNA-Seq approach, the CdMV-cardamom interaction mechanism and associated pathway in mosaic disease have been deciphered and can be used in germplasm improvement against this disease. The complete genetic information for a few cardamom-infecting viruses has been determined but not for bacteria, fungi and other cardamom-infecting organisms. Moreover, cardamom is cultivated under diverse cropping systems encompassing a wide variety of other crops which leads to the transmigration of pathogens from other crops into cardamom causing the emergence of new diseases. Hence a continuous genomics study is needed against the new pathogens for the development of resistant varieties. Finally, adapting genomics in cardamom research would accelerate many different research areas ranging from marker discovery, genetic diversity and linkage/association mapping to genomic selection, physical mapping, gene discovery and genomic-assisted breeding to improve biotic stress resistance in cardamom.

References

Agnihothrudu V (1968) *Phaeodactylium venkatesanum* gen. et sp. nov. on *Elettaria cardamomum* Mat. from Kerala. In: Proceedings of the Indian academy of sciences-section B, vol 68, no 4. Springer India, pp 206–209

Alam A, Rehman NU, Ansari MN, Palla AH (2021) Effects of essential oils of *Elettaria cardamomum* grown in India and Guatemala on gram-negative bacteria and gastrointestinal disorders. Molecules 26(9):2546

Ali MIM (1982) Field evaluation of fungicides against leaf blotch disease of cardamom. Pesticides 11:38–39

Alves MS, Dadalto SP, Gonçalves AB, De Souza GB, Barros VA, Fietto LG (2014) Transcription factor functional protein-protein interactions in plant defense responses. Proteomes 2(1):85–106

Anisha CS, Mathew MK, Sasidharan S, Jose S, Varghese RC, Ranjanan R, Geethu M, Rao YS, Remashree AB (2020) Diversity analysis of released varieties of Indian cardamom using ISSR markers reveal narrowing genetic base

Anjali N, Dharan SS, Nadiya F, Sabu KK (2015) Development of EST-SSR markers to assess genetic diversity in *Elettaria cardamomum* Maton. Int J Appl Sci Biotechnol 3(2):188–192

Anjali N, Ganga KM, Nadiya F, Shefeek S, Sabu KK (2016) Intraspecific variations in cardamom (*Elettaria cardamomum* Maton): assessment of genomic diversity by flow cytometry, cytological studies and ISSR analysis. SpringerPlus 5(1):1–11

Anjali N, Nadiya F, Thomas J, Sabu KK (2019) Identification and characterization of drought responsive microRNAs and their target genes in cardamom (*Elettaria cardamomum* Maton). Plant Growth Regul 87(2):201–216

Anuradha C, Balasubramanian V, Selvarajan R (2015) Sequence motif comparison and homology modeling of helper component proteinase (HC–Pro) of banana bract mosaic virus

Arnao MB, Hernández-Ruiz J (2019) Melatonin and reactive oxygen and nitrogen species: a model for the plant redox network. Melatonin Res 2(3):152–168

Asghar A, Butt MS, Shahid M, Huang Q (2017) Evaluating the antimicrobial potential of green cardamom essential oil focusing on quorum sensing inhibition of *Chromobacterium violaceum*. J Food Sci Technol 54(8):2306–2315

Ausubel FM (2005) Are innate immune signaling pathways in plants and animals conserved? Nat Immunol 6(10):973–979

Ayres PG (1992) Plants versus pests and pathogens: an old story but the same story? In: Environmental plant biology series, pp 1–12

Babu KN, Ravindran PN, Peter KV (1997) Protocol of micropropagation of spices and aromatic crops. Indian Institute of Spices Research, Calicut, p 35

Babu KN, Minoo D, Geetha SP, Samsudeen K, Rema J, Ravindran PN, Peter KV (1998) Plant biotechnology—its role in improvement of spices. Indian J Agric Sci 68(8):533–547

Babu KN, Jayakumar VN, Divakaran M, Venugopal MN, Sudarshan MR, Radhakrishnan VV, Backiyarani S, Narayanaswami MM, Peter KV, Parthasarathy VA (2012) Genetic diversity and phylogenetic relationship among small cardamom (*Elettaria cardamomum* Maton.) cultivars and related genera using DNA markers. Int J Innov Hortic 1(1):47–56

Babu KN, Suraby EJ, Cissin J, Minoo D, Pradeepkumar T, Parthasarathy VA, Peter KV (2013) Status of transgenics in Indian spices. J Trop Agric 51(1):1–14

Backiyarani S, Manohari C, Jebasingh T, Jacob T, Usha R (2007) Cloning of coat protein gene of Kursupara isolate of cardamom mosaic virus for developing transgenic virus-resistant cardamom. In: Recent trends in horticultural biotechnology, vol II. ICAR national symposium on biotechnological interventions for improvement of horticultural crops: issues and strategies, Vellanikkara, Kerala, India, 10–12 Jan 2005. New India Publishing Agency, pp 693–698

Bhat AI, Pamitha NS, Gopika A, Biju CN (2018) Complete genome sequencing of banana bract mosaic virus isolate infecting cardamom revealed its closeness to banana infecting isolate from India. VirusDisease 29(2):212–215

Bhat AI, Pamitha NS, Naveen KP, Biju CN (2020) Identification and characterization of cardamom vein clearing virus, a novel aphid-transmitted nucleorhabdovirus. Eur J Plant Pathol 1–10

Bhai RS, Thomas J, Naidu R (1988) Anthracnose—a new disease of small cardamom. Curr Sci 57(24):1346–1347

Chen L, Wang MR, Li JW, Feng CH, Cui ZH, Zhao L, Wang QC (2019) Exogenous application of melatonin improves eradication of apple stem grooving virus from the infected in vitro shoots by shoot tip culture. Plant Pathol 68(5):997–1006

Cyriac A, Paul R, Anupama K, Sheeja TE, Babu KN, Parthasarathy VA (2016) Isolation and characterization of genomic microsatellite markers for small cardamom (*Elettaria cardamomum* Maton) for utility in genetic diversity analysis. Physiol Mol Biol Plants 22(2):219–229

Debnath B, Islam W, Li M, Sun Y, Lu X, Mitra S, Hussain M, Liu S, Qiu D (2019) Melatonin mediates enhancement of stress tolerance in plants. Int J Mol Sci 20(5):1040

Dingle TC, MacCannell DR (2015) Molecular strain typing and characterisation of toxigenic *Clostridium difficile*. Methods Microbiol 42:329–357

Elangovan S, Jebasingh T (2023) Viral diseases of *Elettaria cardamomum*. In Cardamom (*Elettaria cardamomum*): production, processing and properties (pp 67–80). Springer International Publishing, Cham

Elangovan S, Srikakulam N, Pandi G, Jacob T, Ramakrishnan U, Narayanarao VM, Tennyson J (2019) The first complete genomic sequence of cardamom mosaic virus, a member of the genus Macluravirus (family Potyviridae). Adv Virol 164(6):1723–1726

Gao R, Liu P, Wong SM (2012) Identification of a plant viral RNA genome in the nucleus. PLoS ONE 7(11):e48736

Geetha SP, Babu KN, Rema J, Ravindran PN, Peter KV (2000) Isolation of protoplasts from cardamom (*Elettaria cardamomum* Maton.) and ginger (*Zingiber officinale* Rosc.). J Spices Aromat Crops 9(1):23–30

Hildebrand EM (1958) Masked virus infection in plants. Annu Rev Microbiol 12(1):441–468

Huang X, Yan HD, Zhang XQ, Zhang J, Frazier TP, Huang DJ et al (2016) De novo transcriptome analysis and molecular marker development of two Hemarthria species. Front Plant Sci 7:496

Jacob T, Usha R (2001) 3′-Terminal sequence analysis of the RNA genome of the Indian isolate of cardamom mosaic virus: a new member of genus Macluravirus of Potyviridae. Virus Genes 23(1):81–88

Jebasingh T, Pandaranayaka EP, Mahalakshmi A, Kasin Yadunandam A, Krishnaswamy S, Usha R (2013) Expression, purification and molecular modeling of the NIa protease of cardamom mosaic virus. J Biomol Struct Dyn 31(6):602–611

Jebasingh T, Backiyarani S, Manohari C, Somanath A, Usha R (2017) Transformation of cardamom with the RNA dependent RNA polymerase gene of cardamom mosaic virus. J Appl Biotechnol Bioeng 3(3):00066

Jones-Rhoades MW, Bartel DP, Bartel B (2006) MicroRNAs and their regulatory roles in plants. Annu Rev Plant Biol 57:19–53

Josephrajkumar A, Backiyarani S, Murugan M (2002) PV-1, a pseudostem borer tolerant cardamom variety. Insect Environ 7:155–156

Khan A, George KJ, Jasrotia RS, Aravind S, Angadi UB, Iquebal MA, Manju KP, Jaiswal S, Umadevi P, Rao A, Kumar D (2020) Plant virus interaction mechanism and associated pathways in mosaic disease of small cardamom (*Elettaria cardamomum* Maton) by RNA-Seq approach. Genomics 112(2):2041–2051

Kumar A, Umapathy G, Prabhakar A (1997) A study on the management of fragmented rain forests of the Western Ghats for the conservation of small mammals. Salim Ali Centre for Ornithology and Natural History, Coimbatore, India

Kumar A, Prameela TP, Bhai RS, Siljo A, Biju CN, Anandaraj M, Vinatzer BA (2012) Small cardamom (*Elettaria cardamomum* Maton.) and ginger (*Zingiber officinale* Roxb) bacterial wilt is caused by same strain of *Ralstonia solanacearum*: a result revealed by multilocus sequence typing (MLST). Eur J Plant Pathol 132(4):477–482

Kumar A, Prameela TP, Suseelabhai R (2013) A unique DNA repair and recombination gene (recN) sequence for identification and intraspecific molecular typing of bacterial wilt pathogen *Ralstonia solanacearum* and its comparative analysis with ribosomal DNA sequences. J Biosci 38(2):267–278

Kumar PK, Radhakrishnan VV, Hrideek TK, Sunil S, Kuruvilla KM, Sudharshan MR (2015) Shade trees and forage behaviour of honey bees in cardamom plantations. In: Prospects in forestry and agriculture. Kerala Forest Research Institute KFRI, Peechi, India, pp 112–117

Kumar A, Munjal V, Sheoran N, Prameela TP, Suseelabhai R, Aggarwal R, Jain RK, Eapen SJ (2017) Draft genome sequence of highly virulent race 4/biovar 3 of *Ralstonia solanacearum* CaRs_Mep causing bacterial wilt in Zingiberaceae plants in India. Genome Announc 5(1):e01420-16

Kumar RS, Aravind S, Biju CN, Ankegowda SJ (2018) Characterization of genotypes of small cardamom (*Elettaria cardamomum* Maton) for yield parameters and disease resistance. Indian Soc Plant Crops 46(2):143

Kuriakose G, Sinu PA, Shivanna KR (2009) Domestication of cardamom (*Elettaria cardamomum*) in Western Ghats, India: divergence in productive traits and a shift in major pollinators. Ann Bot 103(5):727–733

Liu SR, Zhou JJ, Hu CG, Wei CL, Zhang JZ (2017) MicroRNA-mediated gene silencing in plant defense and viral counter-defense. Front Microbiol 8:1801

Madhusoodanan KJ (2012) Vistas in breeding cardamom. In: Sasikumar B, Dinesh R, Anandaraj M (eds) The capsule golden jubilee souvenir. Indian Institute of Spices Research, Kozhikode, pp 59–60

Magnaye LV, Espino RRC (1990) Note: banana bract mosaic, a new disease of banana, I. Symptomatology. Philip Agric 73(1):55–59

Maiden MC, Bygraves JA, Feil E, Morelli G, Russell JE, Urwin R, Zhang Q, Zhou J, Zurth K, Caugant DA, Feavers IM, Achtman M, Spratt BG (1998) Multilocus sequence typing: a portable approach to the identification of clones within populations of pathogenic microorganisms. Proc Natl Acad Sci 95(6):3140–3145

Manohari C, Backiyarani S, Jebasingh T, Archana S, Usha R (2008) Efficient plant regeneration in small cardamom (*Elettaria cardamomum* Maton) through somatic embryogenesis

Mathew KM, Reshma R, Geethu M, Rithin V, Sabu KK, Nadiya F, Noushad MA, Dharan SS, Rao YS, Remashree AB (2019) Data on small cardamom transcriptome associated with capsule rot disease. Data Brief 27:104625

Mayne WW (1942) Report on cardamom cultivation in South India. Misc Bull 50:67

Menon MR, Sajoo BV, Ramakrishnan CK, Devi LR (1972) New Phytophthora disease of cardamom. Curr Sci

Michelmore R (2000) Genomic approaches to plant disease resistance. Curr Opin Plant Biol 3(2):125–131

Mishra S, Mitra S, Radhika NS, Yadav A, Rao GP (2019) Multilocus gene-specific characterization of '*Candidatus Phytoplasma australasia*' associated with shoot proliferation disease of small cardamom in India. 3 Biotech 9(11):1–5

Muthappa BN (1965) A new species of Sphaceloma on Cardamon from India. Sydowia 19(1–6):143–145

Nadgauda RS, Mascarenhas AF, Madhusoodanan KJ (1983) Clonal multiplication of cardamom (*Elettaria cardamomum* Maton) by tissue culture. J Plant Crops

Nadiya F, Anjali N, Thomas J, Gangaprasad A, Sabu KK (2019) Deep sequencing identified potential miRNAs involved in defence response, stress and plant growth characteristics of wild genotypes of cardamom. Plant Biol 21(1):3–14

Nair RR (1979) Investigations of fungal diseases of cardamom. Doctoral dissertation, Ph.D. thesis, Kerala Agricultural University, Vellanikkara, Trichur

Nair KP (2020) Cardamom pathology. In: The geography of cardamom (*Elettaria cardamomum* M.). Springer, Cham, pp 125–163

Padmini K, Venugopal MN, Ankegowda SJ (2001) Heterosis for seedling characters in cardamom (*Elettaria cardamomum* Maton). J Spices Aromat Crops 9(2):145–151

Palani SN, Sankaranarayanan R, Tennyson J (2020) Novel interactions of cardamom mosaic virus VPg with cardamom histones H3 and H4. 3 Biotech 10(10):1–5

Park W, Li J, Song R, Messing J, Chen X (2002) CARPEL FACTORY, a dicer homolog, and HEN1, a novel protein, act in microRNA metabolism in *Arabidopsis thaliana*. Curr Biol 12(17):1484–1495

Peter KV, Ravindran PN, Babu NK, Venugopal MN, Geetha SP, Benny D (2001) Production of somaclones and somatic hybrids of cardamom (*Elettaria cardamomum* Maton) for high yield and resistance to diseases. ICAR project report. Indian Institute of Spices Research, Calicut, p 120

Pfeffer S, Zavolan M, Grässer FA, Chien M, Russo JJ, Ju J, John B, Enright AJ, Marks D, Sander C, Tuschl T (2004) Identification of virus-encoded microRNAs. Science 304(5671):734–736

Pireyre M, Burow M (2015) Regulation of MYB and bHLH transcription factors: a glance at the protein level. Mol Plant 8(3):378–388

Prasath D, Venugopal VA (2004) Genetic diversity and conservation of cardamom (*Elettaria cardamomum* Maton.) in India. Plant Genet Resour Newslett 138:55–60

Praveena R, Biju CN, Senthil Kumar R, Darshana CN, Jashmi KC (2013) Preliminary evaluation of cardamom accessions against leaf blight/Chenthal disease. Indian Phytopathol 66:112–113

Pumplin N, Voinnet O (2013) RNA silencing suppression by plant pathogens: defence, counter-defence and counter-counter-defence. Nat Rev Microbiol 11(11):745–760

Rangaswami G, Seshadri VS, Channamma KL (1968) New Cercospora leaf spot of cardamom. Curr Sci

Rao SNK, Narayanaswamy S, Chacko EK, Swamy D (1982) Regeneration of plantlets from callus of *Elettaria cardamomum* Maton

Reiter RJ, Tan DX (2002) Melatonin: an antioxidant in edible plants. Ann N Y Acad Sci 957(1):341–344

Roossinck MJ (2010) Lifestyles of plant viruses. Philos Trans R Soc B Biol Sci 365(1548):1899–1905

Roossinck MJ, Saha P, Wiley GB, Quan J, White JD, Lai H, Chavarria F, Shen G, Roe BA (2010) Ecogenomics: using massively parallel pyrosequencing to understand virus ecology. Mol Ecol 19:81–88

Sakthipriya M, Sabu KK (2018) Development and cross-genera transferability of ginger EST-SSR markers for cardamom. Curr Bioinform 13(1):95–99

Sankaranarayanan R, Palani SN, Kumar A, Punitha Selvakumar AS, Tennyson J (2020) Prediction and experimental confirmation of banana bract mosaic virus encoding miRNAs and their targets. ExRNA 2(1):1–8

Sasikumar B, Dinesh R, Anandaraj M (eds) (2012) The capsule-golden jubilee souvenir, Cardamom Research Centre, Appangala. Indian Institute of Spices Research, Kozhikode

Sharon A, Biju CN, Ankegowda SJ, Shivakumar MS, Kumar RS, Peeran MF (2020) Screening small cardamom (*Elettaria cardamomum* Maton) field gene bank accessions for phenotypic characters, yield potential and disease resistance. J Spices Aromat Crops 29(2)

Sidharthan VK, Kalaivanan NS, Baranwal VK (2021) Identification of two putative novel RNA viruses in the transcriptome datasets of small cardamom. Plant Gene 100305

Siljo A, Bhat AI, Biju CN, Venugopal MN (2012) Occurrence of banana bract mosaic virus on cardamom. Phytoparasitica 40(1):77–85

Sridhar VV, Muthusamy M, Naidu R (1990) Survey on the occurrence of Nilgiri necrosis—a new virus disease on cardamom. South Indian Horticult 38(3)

Sun G (2012) MicroRNAs and their diverse functions in plants. Plant Mol Biol 80(1):17–36

Thirumalachar MJ (1943) A new rust disease of cardamoms. Curr Sci 12(8)

Thomas J (2000) Biological control of rot diseases of small cardamom. In: Biocontrol potential and its exploitation in sustainable agriculture. Springer, Boston, MA, pp 223–237

Thomas J (2001) Biological control of insect pests of small cardamom. In: Biocontrol potential and its exploitation in sustainable agriculture. Springer, Boston, MA, pp 389–399

Venugopal MN (1999) Natural disease escapes as source of resistance against cardamom mosaic virus causing katte disease of cardamom (*Elettaria cardamomum* Maton). J Spices Aromat Crops 8:145–151

Venugopal MN (2002) Viral diseases of cardamom. In: Cardamom. CRC Press, pp 159–175

Venugopal MN, Prasath D, Mulge R (2006) IISR Avinash—a rhizome rot resistant and high yielding variety of cardamom (*Elettaria cardamomum* Maton). J Spices Aromat Crops 15(1):14–18

Verma S, Nizam S, Verma PK (2013) Biotic and abiotic stress signaling in plants. In: Stress signaling in plants: genomics and proteomics perspective, vol 1. Springer, New York, NY, pp 25–49

Wang IC, Sether DM, Melzer MJ, Borth WB, Hu JS (2010) First report of banana bract mosaic virus in flowering ginger in Hawaii. Plant Dis 94(7):921

Zhao L, Chen L, Gu P, Zhan X, Zhang Y, Hou C, Wu Z, Wu YF, Wang QC (2019) Exogenous application of melatonin improves plant resistance to virus infection. Plant Pathol 68(7):1287–1295

Genomics of Flavor Constituents of Cardamom

6

R. Sivaranjani

Abstract

Cardamom, scientifically known as *Elettaria cardamomum* (L.) Maton, is esteemed as the queen of spices, cherished globally for its distinctive flavor. This chapter delves into the intricate genomics underlying the flavor constituents of cardamom, shedding light on its biochemical composition, aroma profile, and medicinal properties. Variations in these constituents across different genotypes, environmental conditions, and stages of maturity are elucidated, providing insights into the factors shaping cardamom's flavor profile. Beyond its culinary significance, cardamom holds a revered place in traditional medicine, with documented antimicrobial, anti-inflammatory, and gastro-protective properties. The biosynthesis of terpenoids, the primary constituents of cardamom essential oil, is elucidated, offering a comprehensive overview of the biochemical pathways involved. Advances in genomic research, including the identification of partial gene sequences and SSR markers, offer promising avenues for enhancing cardamom's flavor profile through targeted breeding and genetic modification. Ultimately, the convergence of genomic insights and biotechnological innovations holds immense potential for unlocking the full flavor potential of cardamom, paving the way for engineered solutions to meet evolving culinary and medicinal needs.

Keywords

Cardamom · Genomics · Flavor constituents · Essential oil · Terpenoid biosynthesis

6.1 Introduction

The queen of spice, Cardamom scientifically known as *Elettaria cardamomum* (L.) Maton belongs to the family Zingiberaceae. It is an indispensable spice used in different cuisines worldwide and is the only spice preferred exclusively for its flavor. Cardamom seeds and fruits are the economic produce obtained from the plant. The volatile oil constituents concentrated mainly on the seeds of the cardamom capsule with only minuscule quantity contributed by the outer husk. The aroma of the cardamom enhances the sensory quality of the food preparations. Due to its sweet aroma, it is mainly used in the preparation of sweet dishes such as desserts, drinks, ice creams, cool drinks, candies, chewing gum, biscuits, and other baked food

R. Sivaranjani (✉)
ICAR-Indian Institute of Spices Research, Kozhikode, Kerala, India
e-mail: Sivaranjani.R@icar.gov.in

K. K. Sabu and C. Kole (eds.), *The Cardamom Genome*, Compendium of Plant Genomes,
https://doi.org/10.1007/978-3-031-71829-8_6

products. In Arabic countries, it is also used to flavor tea and coffee beverages. In India and Indonesia, the cardamom seeds are used along with betel leaves as mouth fresheners. India exports the volatile oil extracted from cardamom seeds and it stands first in the global export of cardamom essential oil. The USA, Germany, Brazil, and Australia are the biggest importer of Indian cardamom essential oil. As per FEMA GRAS assessment report of natural flavors, the estimated intake of cardamom seed essential oil is 340 μg/day/person (Fukushima et al. 2020).

The nutrient composition of cardamom pods revealed that it is rich in dietary fiber (28%), calcium (383 mg/100 g), magnesium (229 mg/100 g), potassium (1119 mg/100 g), and both saturated (myristic, palmitic, and stearic acid) and unsaturated fatty acids (oleic, linoleic, and linolenic) and phytosterols (USDA 2012). The percentage of cardamom seed essential oil varies from 5 to 9% (v/v) among different genotypes. The essential oil contains oxygenated monoterpenes in higher proportion (> 80%) with around 8–10% of monoterpene hydrocarbons. The oxygenated monoterpenes include linalool, 1,8-cineole, terpinen-4-ol, α-terpineol, Geraniol, α-terpinyl acetate, linalyl acetate, neryl acetate, and monoterpene hydrocarbon pool includes, thujene, sabinene, α-pinene, β-pinene, myrcene, and limonene. The volatile constituents in the cardamom seeds vary with genotypes, climate, soil and environmental condition, and stages of maturity (Alagupalamuthirsolai et al. 2020).

In a study, 22 cardamom accessions collected from southern India were analyzed for the presence of volatile constituents in the essential oil extracted from seeds of cardamom. The variation in the constituents reported in the study is thujene (0.1–0.9%), sabinene (1.9–4.9%), α-pinene (0.6–1.5%), myrcene (1.0–1.9%) and limonene (0.9–9.4%), linalool (0.4–11.0%), 1,8-cineole (15.2–49.4%), α-terpineol (0.8–13.2%), α-terpinyl acetate (29.9–61.3%), linalyl acetate (2.2–4.4%) (Ashokkumar et al. 2021).

Cardamom has been classified into Malabar, Mysore, and Vazhukka types based on its habitat adaptability and growth characteristics. Phytochemical investigations revealed that Mysore-type cardamom variety such as Alleppey Green is superior in flavor quality as they contain low amount of 1,8-cineole and higher amount of α-terpinyl acetate, linalool, and linalyl acetate (Nair 2020). Moreover, this variety has given higher α-terpinyl acetate content over five seasons and gave higher essential oil yield. Phytochemical profiling showed that Guatemalan and Ceylon cardamom is superior marginally to Indian cardamom due to the higher content of α-terpinyl acetate (Pillai et al. 1984).

Cardamom essential oil has a spicy-sweet middle note contributed by the above-mentioned terpenoid compounds. The major volatile constituents and their aroma notes are explained in Table 6.1.

Though most of the compounds identified in cardamom are found in many other aromatic crops, combination of 1,8-cineole and α-terpinyl acetate in major proportion makes its flavor unique. The phytochemical polymorphism where differences in the chemical composition of volatile oil in different plant organs is a common case in most of the aromatic plants including cardamom. Jena et al. (2021) reported that cardamom leaf volatile oil contains 1,8-cineole (20.7%) as the primary compound followed by comphene (18.1%), camphor (10.0%) and tricyclene (7.3%) whereas the volatile oil extracted from stem contains α-terpinyl acetate (19.8%), 1,8-cineole (10.4%), caryophyllene oxide (7.1%) and β-eudesmol (4.9%).

The volatile constituents belonging to monoterpenes and sesquiterpenes are termed as secondary metabolites without having direct role in plant growth and regulation but recent studies showed that these compounds play very important ecological role in plant's interaction with their environment including acting as pollinating and allelopathic agents.

Table 6.1 The aroma notes of volatile constituents of cardamom essential oil

S. No.	Compound name	Aroma notes of the volatile compound
1	α-pinene	Warm, refreshing, herbal, woody, and slightly camphoraceous with a cooling effect. A refreshing top note
2	Sabinene	Spicy, pine-like scent with notes of citrus
3	β-myrcene	Earthy, fruity, and clove-like
4	1,8-cineole	Camphoraceous and slightly woody notes
5	ϒ-terpinene	Characteristic lemon odor
6	Sabinene hydrate	Citrus and pine
7	α-terpinolene	Citrus-y, woody, terpy with camphoraceous and thymol notes
8	Linalool	Intense floral, lavender and sweet notes
9	Terpinen-4-ol	Warm, peppery, mildly earthy, musty-woody odor of moderate tenacity
10	α-terpineol	Citrus-woody profile and lemon and lime nuances
11	Linalyl acetate	Floral, sweet and citric, and additionally as minty and slightly caraway-like
12	Geraniol	Sweet floral rose-like, citrus with fruity, waxy nuances
13	α-Terpinyl acetate	Mildly herbaceous, sweet and refreshing odor of spicy bergamot-lavender type, with variations into piney note
14	Neryl acetate	Fresh floral, citrus white rose-like aroma

6.2 Traditional Medicinal Uses

Cardamom is used as an important ingredient in several traditional systems of medicines including Ayurveda and Unani. It has antimicrobial, analgesic, anti-inflammatory, diuretic, and stimulant properties, which has been used for the treatment of disorders of gastro-intestinal tract such as constipation, diarrhoea, colic infections, consumption, etc. along with other ailments such as epilepsy, lung congestion, bronchitis, bladder, and kidney-related complications (Korikanthimathm et al. 2000).

6.3 Medicinal Properties of Cardamom Essential Oil

Cardamom has been reported to have several pharmacological properties such as gastroprotective, antihypertensive, anticancer, anti-inflammatory, antimicrobial, and antiplatelet aggregation activities as proved by various in vitro assays (Abdullah et al. 2022). It is also used in the prevention and cure of digestive disorders, pulmonary tuberculosis, and throat infections. The volatile essential oil extracted from cardamom seeds is used in healthy foods, beverages, and also in perfume preparations (Kaur et al. 2013). Alcohol extract of cardamom, ameliorated the symptoms of induced type 2 diabetes in rats through alleviation of oxidative stress and activation of insulin signaling pathway in brain. Cardamom oil also possesses antidiabetic, cardioprotective, and hypolipimidic activities as proved by several in vivo studies (Gomaa et al. 2019; Goyal et al. 2015; Ahmed et al. 2024).

Alam et al. (2021) showed that essential oil obtained from Indian cardamom had more potent antidiarrheal and antispasmodic effects by inhibiting the growth of *P. aeruginosa* and *E. coli* efficiently than Guatemala-sourced cardamom. The reason they cited is due to the presence of high amount of 1,8-cineole in Indian cardamom than its Guatemalan counterparts which contains high α-terpinyl acetate content.

They also proved that cardamom essential oil possesses good antioxidant, anti-inflammatory activities via several in vitro and in silico molecular docking analysis of volatile constituents of cardamom.

Chowdhury and Kumar (2020) reported through in vitro analysis that α-terpinyl acetate inhibited acetylcholinesterase and butyrylcholinesterase, target enzymes for Alzheimer's disease (AD) treatment. They also reported that this compound reduced Aβ-induced neurotoxicity, hydrogen peroxide induced oxidative stress, and possess anti-amyloidogenic property. They concluded that α-terpinyl acetate has multi-target potential and disease amelioration effects in AD model. Not only α-terpinyl acetate, but α-terpineol also reported to have anti-Alzheimer's activity proved in rodent *in vivo* models. In preventive and therapeutic modes, α-terpineol consumption could improve neurogenesis and long-term memory while reducing amyloid plaque counts and ameliorating biochemical factors responsible for plague formation (higher levels of superoxide dismutase and malondialdehyde and reduced levels of MDA) (Alipour et al. 2022).

1,8-cineole, also known as eucalyptol is reported to have various pharmacological properties such as bronchodilator, mycolytic, anti-inflammatory, antimicrobial, analgesic, antioxidant, neuroprotective, expectorant, antispasmodic, etc. (Juergens 2014; Santos and Rao 2000). A clinical trial by Worth et al. (2009) investigated the effects of 1,8-cineole in patients with chronic obstructive pulmonary disease (COPD) through randomized, double-blind, placebo-controlled study. The authors concluded that concomitant therapy with 1,8-cineole improves lung function and health status of the patients. This study further suggests that cineole intervenes in the pathophysiology and controls the inflammation of the mucus membrane. Paul et al. (2019) showed through several in vitro assays that 1,8-cinole-rich cardamom extract prepared through supercritical carbon dioxide extraction showed better antidiabetic activity than pure 1,8-cinole.

The fractions rich in α-terpineol and α-terpinyl acetate obtained from acetone extract of cardamom seed was reported to enhance the skin penetration capacity of prednisolone drug as compared to chemical enhancer Azone in mouse model (Yamahara et al. 1989). In yet another work by Huang et al. (1996), 5% cardamom volatile oil increased the plasma concentration of piroxicam drug (67.09 fold) compared to control when injected through rabbit abdominal skin. The essential oil also increased the bioavailability of the drug to 83.23%.

6.4 Toxicity

The cardamom essential oil was tested negative for its mutagenicity with different strains of *S. typhimurium* and it was negative in an Unscheduled DNA Synthesis (UDS) assay in rat hepatocytes at a dose of 50 mg/mL. However, some developmental toxicity with clinical symptoms such as salivation, decreased motor function, and neurological conditions such as twitches and tremors are observed at dose levels of 375, 750, or 1500 mg/kg bw/day administered for seven days prior to cohabitation, gestation, delivery, and a four-day lactation/post-parturition period in female rats. Based on the lack of adverse effects of cardamom oil on mating, fertility, duration of gestation, or duration of parturition, a no-observed-adverse-effect-level (NOAEL) of 1500 mg/kg bw/day was determined for reproductive effects. The NOAEL for the offspring was determined to be 375 mg/kg bw/day based on the decreased body weights (Hoberman 1989).

6.5 Biosynthesis of Terpenoids

The volatile oil from cardamom seeds primarily contains monoterpene hydrocarbons and oxygenated monoterpenes as mentioned in the previous section. Monoterpenes are C10 compounds primarily synthesized in plastids. They are highly volatile compounds having low boiling points as compared to sesquiterpene compounds, which are C15 terpenoid compounds synthesized primarily in the cytosol. All terpenoids are biosynthesized from repeated addition of 4-carbon isoprene unit in a series of enzyme-catalyzed reaction.

6.5.1 Synthesis of IPP and DMAPP

The substrates for the biosynthesis of the terpenoids are two 4C isoprene units such as isopentenyl pyrophosphate (IPP) and dimethylallylpyrophosphate (DMAPP). DMAPP is the isomerized version of IPP catalyzed by the isopentyl pyrophosphate isomerase (IDI). The IPP and DMAPP are synthesized in plastids via methylerythritol 4-phosphate (MEP) pathway which leads to the production of monoterpenoids. The pathway for triterpenoid and sesquiterpenoid biosynthesis occurs primarily in the cytosol but also in the peroxisome, in this case, the IPP and DMAPP are synthesized through mevalonate (MVA) biosynthetic pathway. Figures 6.1 and 6.2 explain the detailed flow diagram of synthesis of IPP and DMAPP via MEP and MVA biosynthetic pathways.

MVA pathway uses acetyl CoA whereas MEP pathway uses pyruvate and glyceraldehyde-3-phosphate as substrates for the synthesis of IDP and DMAPP.

HMG-CoA reductase (HMGR) in the MVA pathway was identified as a rate-limiting enzyme by Bach (1986). In later years, numerous studies involving inhibition and overexpression of HMGR in different model plants led to a better understanding of this enzyme in the increased production of terpenoid compounds. Besides this enzyme, HMG-CoA synthase (HMGS) is also considered an effective target for genetic manipulation for increased terpenoid production. In the MEP pathway, DXS is shown as the first rate-limiting step. The manipulation of this enzyme has been demonstrated for improved production of IPP, DMAPP, and downstream terpenoids in different species such as *Synechococcus leopoliensis*, *Lycopersicon esculentum,* and *Streptomyces*. For effective production of increased isoprene, combinatorial approach like directed co-evolution of key enzymes such as HMGR, DXS, and promotor modification is required (Wang et al. 2019).

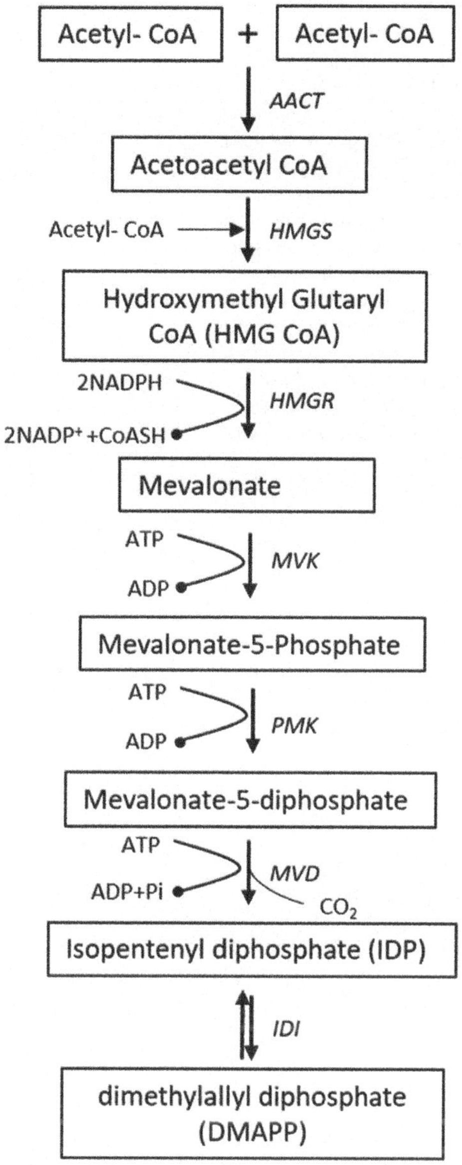

Fig. 6.1 Biosynthesis of isopentenyl diphosphate (IDP) and dimethylallyl diphosphate (DMADP) via the mevalonic acid (MVA) pathway (*AACT*—Acetoacetyl-CoA thiolase; *HMGS*—hydorxymethylglutaryl-CoA synthase; *HMGR*—hydroxymethylglutaryl-CoA reductase; *MVK*—mevalonate kinase; *PMK*—phosphomevalonate kinase; *MVD*—mevalonate diphosphate decarboxylase; *IDI*—isopentenyl diphosphate isomerase)

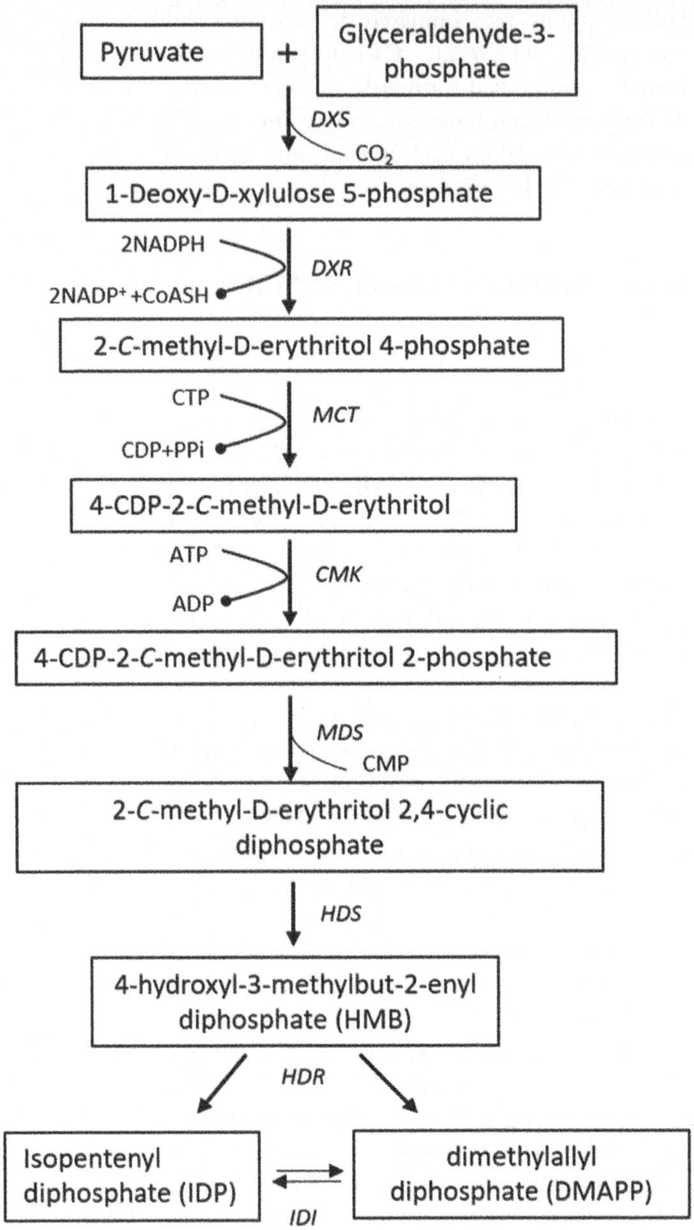

Fig. 6.2 Biosynthesis of isopentenyl (IDP) and dimethylallyl diphosphate (DMADP) via the MEP pathway (*DXS*—1-Deoxy-D-xylulose 5-phosphate synthase; *DXR*—1-Deoxy-D-xylulose 5-phosphate reductoisomerase; *MCT*—2-C-methyl-D-erythritol 4-phosphate cytidylyltransferase; *CMK*—4-diphosphocytidyl-2-C-methyl-D-erythritol kinase; *MDS*—2-C-methyl-D-erythritol 2,4-cyclodiphosphate synthase; *HDS*—HMB-PP synthase; *HDR*—HMB-PP reductase; *IDI*—Isopentenyl diphosphate isomerase)

6.5.2 Synthesis of GPP

Once synthesized, IPP and DMAPP are used as feed material for the synthesis of different classes of terpenoids. The condensation of IPP and DMAPP results in the synthesis of geranyl diphosphate (GPP) catalyzed by geranyl diphosphate synthase (GDPS). GPP is a C10 compound, which forms the base for the synthesis of other monoterpene compounds. Secondary metabolites synthetic pathway always relies primary pathway for its substrates, precursors, energy source (ATP), and co-factors. This relationship explains the varied yield and accumulation of essential oil in different crops with respect to factors affecting the growth and development of the crop such as soil, climatic factors, nutrient application, pest and disease incidence, etc.

6.5.3 Catalytic Conversion of GPP to Monoterpenes

The conversion of GPP to monoterpenes is catalyzed by different monoterpene synthases according to plant species. In cardamom, this reaction is catalyzed by cineole synthase which cyclize the GPP into 1,8-cineole. Most of the monoterpene synthases including cineole synthase have higher activity in the plastids, though some are shown to be active in the mitochondria and cytosol too (Dong et al. 2013). A study by Dong et al. (2015) showed that GPP is exchanged between different subcellular organelles as feed material for terpenoid biosynthesis. Not only GPP, but also IPP is also transported from plastids to mitochondria for the synthesis of monoterpenes.

Monoterpene synthases including cineole synthase catalyze the multi-step complex biochemical reactions wherein carbon atoms in the substrate are modified extensively through bonding and stereoselectivity resulted in the synthesis of variety of products. For example, Peters and Croteau (2003) highlighted that in *Saliva officinalis*, GPP undergoes ionization and isomerization to produce 3*R*-linalyl cation and subsequent ionization and cyclization to 4*R*-α-terpinyl cation intermediate. These reactions are common to all monoterpene synthases. In case of 1,8-cineole synthase, this cation intermediate yields limonene and 4*R*-α-terpineol as intermediate leads up to 1,8-cineole by protonation and cyclization. The 2,7-ring closure of 4*R*-α-terpinyl cation leads to the formation of pinyl cation, which gets deprotonated to β-pinene and α-pinene. The 1,8-cineole is a biosynthetic endpoint in many plant systems, the reason for the higher percentage of this compound in cardamom.

In Tea Tree (*Melaleuca alternifolia*), Padovan et al. (2017) isolated and characterized through transcriptomic approach three distinct monoterpene synthases such as 1,8-cineole synthases, terpinolene synthase, sabinene hydrate synthase which produced 1,8-cineole, terpinolene and sabinene hydrate as the prominent monoterpene compounds in its essential oil. The abundance of particular transcript decides the chemotypes of a species. Chemotypes are described as subspecies having same morphological characters but produce different quantities of chemical components in its essential oil (Crocoll et al. 2010).

The other significant monoterpene cyclase includes limonene synthase which catalyzes the cyclization of GPP to (−)-4S-limonene (Croteau 1991, 1993), and sabinene synthase which catalyzes the ionization and isomerization of GPP to 3R-linalyl pyrophosphate. Further ionization and cyclization results in the formation of sabinene with the formation of α-terpinyl cation, the same intermediate involved in the production of 1,8-cineole via α-terpineol.

The biosynthesis of 1,8-cineole in cardamom is believed to be consistent with the biosynthetic pathway of monoterpenes in other plant species. Several transcript encoding cineole synthase from cardamom are reported by Nadiya et al. (2017) through genome-wide transcriptomic analysis. They annotated the assembled transcripts to NCBI non-redundant (NR) protein database, Swissprot, TrEMBL, pfam, PlantCyc, KOG, and TAIR10 databases and identified the transcripts. The annotation of unigenes identified through transcriptomic analysis through PlantCyc enzyme database matched several genes to enzymes involved in secondary metabolite pathways such as phenylpropanoid, isoprenoid, lignin, and alkaloid biosynthetic pathways. The authors identified several transcript encoding 1,8-cineole synthase and also limonene synthase and limonene-6-hydroxylase which are involved in limonene biosynthesis and limonone degradation, respectively.

The same group of authors also validated the expression profile of cineole synthase in wild and cultivated varieties of cardamom. The analysis showed that most of the terpenoid biosynthetic pathway genes showed downregulation in the cultivated variety as compared to wild variety (Nadiya et al. 2018). They found the expression of limonene synthase in wild variety is more when compared to cultivated variety and the same has been proved through volatile analysis of essential oil where the content of limonene was six times higher in wild than cultivated genotype. The other unigenes identified through their study which involved in monoterpene biosynthetic pathway include geranyl-diphosphate diphosphate lyase and borneol dehydrogenase along with several enzymes involved in mevalonate biosynthetic pathway of synthesis of IPP and DMAPP. The identified transcript could be used in the synthesis of full-length cDNA of particular gene which could be used for the development of transgenic plants to increase the accumulation of flavor compounds in the seeds of cardamom in the future. For example, limonene synthase gene from *Perilla frutescens* was transgenically expressed in *Eucalyptus camaldulensis* resulting in 2.6 to 4.5 times increased accumulation of limonene in transgenic plants as compared to wild type (Ohara et al. 2010). The limonene synthase gene, 4S-LS from *Mentha spicata* was transformed to *M. x piperita* and *M. arvensis* to study the effect on the production of terpenoids by Diemer et al. (2001). They showed no significant changes in the accumulation of limonene in the transgenic plants as compared to control, but the amount of pinene showed 56% higher in some transformed plants and 54% lower in some plants. Likewise, in *M. arvensis* transformed plants, the amount of myrcene was found lower than control. The variation in the monoterpene content in the transgenic plants was the result of competition between 4S-LS limonene synthase enzyme and other monoterpene synthase which use GPP as a substrate. These studies reveals the complexity of the biosynthetic pathway of monoterpene synthesis.

Ali et al. (2022) demonstrated in tobacco plants, overexpression of *SgCINS*, cineole synthase gene from *Saliva guaranitica* leaves leads to increased synthesis of linoleic, linolenic esters, phytol, and ledol in transgenic plants as compared to control plants. In transgenic plants, the number of monoterpenes was found higher than control but the sesquiterpenes were found lower than control. Moreover, they established that transgenic tobacco plant with higher monoterpenes content was found highly resistant to cotton worm attacks than control plants.

To date, no full-length sequence of cineole synthase or any genes of cardamom monoterpenoid biosynthetic pathway have been identified. But, in related species of cardamom, Wang et al. (2018) identified terpene synthase gene *AvTPS3* in *Amomum villosum* which has an important role in the biosynthesis of bornyl acetate as a major product along with campene, myrcene, and limonene, which is also present in cardamom. They reported the tissue-specific expression of terpenoid synthase genes. The expression of *AvTPS3* was higher in seeds where accumulation camphene, limonene, β-myrcene, borneol, camphor, and bornyl acetate was higher than leaves. They also reported other gene *AvTPS1* that the authors renamed as *AvPS* (Pinene synthase) catalyzes the synthesis of α-pinene and β-pinene using GPP as substrate but did not form products with FPP (Farnesyl diphosphate).

6.5.4 Biosynthesis of Terpinyl Esters

The main monoterpene ester in cardamom volatile oil, α-terpinyl acetate is synthesized from α-terpineol by trans-esterification reaction but the pathway is not yet fully explored. There is no report on the enzyme involved in the biosynthesis of α-terpinyl acetate in the cardamom per se. Croteau and Hooper (1978) reported the activity of a monoterpenol O-acetyltransferase from *Mentha piperita*. The transacylase enzyme catalyzes the acetylation of menthol to produce menthol acetate. The authors characterized the enzymes and found that it was highly specific for acetyl CoA and were not efficient with propionyl CoA and butyryl CoA as acyl donor. At the same time, the isolated enzyme is not selective with regard to menthol as alcohol substrate and it can utilize other monoterpene alcohol as its substrate too.

Zaks et al. (2008) reported the activity of linalool acetyl transferase (LAT) in a similar line to earlier mentioned report. This enzyme also utilizes acetyl CoA as ester group donor and uses monoterpene alcohol as its substrate. When the authors incubated cell-free extracts of glandular trichomes isolated from lemon mint with linalool and acetyl CoA, linalyl acetate are generated.

The draft whole genome sequence of Njallani Green Gold, ruling variety in cardamom growing tract of Kerala and Karnataka was published by a group of scientists from ICAR-National Bureau of Plant Genetic Resources, New Delhi (Gaikwad et al. 2023). They estimated the genome size to be 1.06 Gb (gigabases). The authors identified more than two lakhs SSR markers from the genome sequence and among them, 246 SSR loci were validated at lab level. The identified SSR markers could be utilized as trait-associated marker in marker-assisted breeding. More information from this draft genome is expected in the coming year.

6.6 Conclusion

The advances in the genomic research of cardamom have led to the identification of partial sequence of major genes involved in the synthesis of flavor compounds in the past decades though the research is still in rudimentary stage. Identification of full-length genes involved in the monoterepene biosynthetic pathway along with its regulatory sequences is a prerequisite in improving the flavor profile of the crop using next generation genomic technologies such as CRISPR/Cas9. The variation in the volatile profile of already existing germplasm resources would assist in the selection of efficient genes with better flavor profiles in the cardamom seeds. The identification of SSR markers from the draft genome should be effectively used in the identification of genotype with higher α-terpinyl acetate and lower 1,8-cineole content. It must be noted that the flavor profile of the cardamom is defined by many other volatile compounds, so future genomic research should focus its effort on the overall improvement of flavor compounds instead of targeting a few compounds. Once the target enzymes are identified, transgenic approaches could be employed to achieve the optimal concentration of different volatile constituents with enhanced flavor profiles in the cardamom. Moreover, the genomic information available in the related species could also be employed for genomic improvement of cardamom.

To conclude, the improvement in the genomic research targeting genes involved in volatile compounds synthesis in cardamom would definitely open the flood-gate to make engineered microbes/model organisms to synthesize the valuable natural flavor compound.

References

Abdullah AN, Tian W, Zengliu S, Zou Y, Farooq S, Huang Q, Xiao J (2022) Recent advances in the extraction, chemical composition, therapeutic potential, and delivery of cardamom phytochemicals. Front Nutr 9:1024820. https://doi.org/10.3389/fnut.2022.1024820

Ahmed VA, Ahmed BS, Aziz TA (2024) Cardioprotective and hypolipidemic effect of cardamom oil-loaded lipid carrier nanoparticles in a rat model of streptozotocin-induced diabetes. Al-Rafidain J Med Sci 6(1):105–111. https://doi.org/10.54133/ajms.v6i1.498

Alagupalamuthirsolai M, Sivaranjani R, Ankegowda SJ, Murugan M, Krishnamurthy KS, Aravind S, Asangi H (2020) A study of essential oil constituents from capsules, physiological and quality parameters of small cardamom ecotypes [*Elettaria cardamomum* (L.) Maton] growing in the Western Ghats, India. J Essent Oil Bear Plants 23(6):1253–1264. https://doi.org/10.1080/0972060X.2020.1869105

Alam A, Jawaid T, Alam P (2021) *In vitro* antioxidant and anti-inflammatory activities of green cardamom essential oil and *in silico* molecular docking of its major bioactives. J Taibah Univ Sci 15(1):757–768. https://doi.org/10.1080/16583655.2021.2002550

Ali M, Alshehri D, AlKhaibari AB, Elhalem NA, Darwish DBE (2022) Cloning and characterization of 1,8-cineole synthase (SgCINS) gene from the leaves of *Salvia guaranitica* plant. Front Plant Sci 13. https://doi.org/10.3389/fpls.2022.869432

Alipour H, Yaghmaei P, Ahmadian S, Ghobeh M, Ebrahim-Habibi A (2022) A study on alpha-terpineol in Alzheimer's disease with the use of rodent in vivo model, restraint stress effect and in vitro amyloid beta fibrils. Braz J Pharm Sci 58:e19090. https://doi.org/10.1590/s2175-97902022e19090

Ashokkumar K, Vellaikumar S, Murugan M, Dhanya MK, Ariharasutharsan G, Aiswarya S, Akilan M,

Warkentin TD, Karthikeyan A (2021) Essential oil profile diversity in cardamom accessions from Southern India. Front Sustain Food Syst 5. https://doi.org/10.3389/fsufs.2021.639619

Bach TJ (1986) Hydroxymethylglutaryl-CoA reductase, a key enzyme in phytosterol synthesis? Lipids 21:82–88

Chowdhury S, Kumar S (2020) Alpha-terpinyl acetate: a natural monoterpenoid from *Elettaria cardamomum* as multi-target directed ligand in Alzheimer's disease. J Funct Foods 68:103892. https://doi.org/10.1016/j.jff.2020.103892

Crocoll C, Asbach J, Novak J, Gershenzon J, Degenhardt J (2010) Terpene synthases of oregano (*Origanum vulgare* L.) and their roles in the pathway and regulation of terpene biosynthesis. Plant Mol Biol 73:587–603. https://doi.org/10.1007/s11103-010-9636-1

Croteau R (1991) Metabolism of monoterpenes in mint (Mentha) species. Planta Med 57(Suppl):10–14

Croteau R (1993) Biosynthesis of limonene in Mentha species. In Schreier P, Winterhalter P (eds) Progress in flavor precursor studies: analysis, generation, biotechnology. Allured Publishing Co, Wheaton, IL, pp 113–122

Croteau R, Hooper CL (1978) Metabolism of monoterpenes: acetylation of (−)-menthol by a soluble enzyme preparation from peppermint (*Mentha piperita*) leaves. Plant Physiol 61(5):737–742. https://doi.org/10.1104/pp.61.5.737

Diemer F, Caissard J-C, Moja S, Chalchat J-C, Jullien F (2001) Altered monoterpene composition in transgenic mint following the introduction of 4S-limonene synthase. Plant Physiol Biochem 39(7–8):603–614. https://doi.org/10.1016/s0981-9428(01)01273-6

Dong L, Miettinen K, Goedbloed M, Verstappen FW, Voster A, Jongsma MA, Memelink J, Krol AVD, Bouwmeester HJ (2013) Characterization of two geraniol synthases from *Valeriana officinalis* and *Lippia dulcis*: similar activity but difference in subcellular localization. Metab Eng 20:198–211

Dong L, Jongedijk E, Bouwmeester H, Van Der Krol A (2015) Monoterpene biosynthesis potential of plant subcellular compartments. New Phytol 209(2):679–690. https://doi.org/10.1111/nph.13629

Fukushima S, Cohen SM, Eisenbrand G, Gooderham NJ, Guengerich FP, Hecht SS, Rietjens IMCM, Rosol TJ, Davidsen JM, Harman CL, Lu V, Taylor SV (2020) FEMA GRAS assessment of natural flavor complexes: lavender, guaiac coriander-derived and related flavoring ingredients. Food Chem Toxicol 11584. https://doi.org/10.1016/j.fct.2020.111584

Gaikwad AB, Kumari R, Yadav S, Rangan P, Wankhede DP, Bhat KV (2023) Small cardamom genome: development and utilization of microsatellite markers from a draft genome sequence of *Elettaria cardamomum* Maton. Front Plant Sci 14. https://doi.org/10.3389/fpls.2023.1161499

Gomaa AA, Makboul RM, El-Mokhtar MA, Abdel-Rahman EA, Ahmed IA, Nicola MA (2019) Terpenoid-rich *Elettaria cardamomum* extract prevents Alzheimer-like alterations induced in diabetic rats via inhibition of GSK3β activity, oxidative stress and pro-inflammatory cytokines. Cytokine 113:405–416

Goyal SN, Sharma C, Mahajan UB, Patil CR, Agrawal YO, Kumari S, Arya DS, Ojha S (2015) Protective effects of cardamom in isoproterenol-induced myocardial infarction in rats. Int J Mol Sci 16(11):27457–27469. https://doi.org/10.3390/ijms161126040

Hoberman AM (1989) Reproductive and developmental toxicity screening test of B9 (cardamom oil) administered orally via gavage to Crl:CD (SD) BR female rats. Unpublished study report, study no. 412-004

Huang Y-B, Wu P-C, Ko H-M, Tsai Y-H (1996) Effect of pretreatment by cardamom oil on *in vivo* percutaneous absorption of piroxicam gel: pharmacokinetic analysis in rabbits. Int J Pharm 134:183–191. https://doi.org/10.1016/0378-5173(95)04430-2

Jena S, Ray A, Sahoo A, Champati BB, Padhiari BM, Dash B, Nayak S, Panda PC (2021) Chemical composition and antioxidant activities of essential oil from leaf and stem of *Elettaria cardamomum* from Eastern India. J Essent Oil Bear Plants 24(3):538–546

Juergens UR (2014) Anti-inflammatory properties of the monoterpene 1.8-cineole: current evidence for co-medication in inflammatory airway diseases. Drug Res (Stuttg) 64(12):638–646. https://doi.org/10.1055/s-0034-1372609

Kaur J, Kaur S, Mahajan A (2013) Herbal medicines: possible risks and benefits. Am J Phytomed Clin Therapeut 2:226–239

Korikanthimathm VS, Prasath D, Rao G (2000) Medicinal properties of cardamom *Elettaria cardamomum*. J Med Aromat Plant Sci 22:683–685

Nadiya F, Anjali N, Thomas J, Gangaprasad A, Sabu KK (2017) Transcriptome profiling of *Elettaria cardamomum* (L.) Maton (small cardamom). Genom Data 11:102–103. https://doi.org/10.1016/j.gdata.2016.12.013

Nadiya F, Anjali N, Thomas J, Gangaprasad A, Sabu KK (2018) Genome-wide differential expression profiling in wild and cultivar genotypes of cardamom reveals regulation of key pathways in plant growth and development. Agri Gene 8:18–27. https://doi.org/10.1016/j.aggene.2018.03.002

Nair KP (2020) The geography of cardamom (*Elettaria cardamomum* M.): the "queen" of spices—volume 2. In: Nair KP (ed), 1st edn. Springer International Publishing: Imprint: Springer

Ohara K, Matsunaga E, Nanto K, Yamamoto K, Sasaki K, Ebinuma H, Yazaki K (2010) Monoterpene engineering in a woody plant *Eucalyptus camaldulensis* using a limonene synthase cDNA. Plant Biotech J 8:28–37. https://doi.org/10.1111/j.1467-7652.2009.00461.x

Padovan A, Keszei A, Hassan Y, Krause ST, Köllner TG, Degenhardt J, Gershenzon J, Külheim C, Foley WJ (2017) Four terpene synthases contribute to the generation of chemotypes in tea tree (*Melaleuca alternifolia*). BMC Plant Biol 17(1):160. https://doi.org/10.1186/s12870-017-1107-2

Paul K, Bhattacharjee P, Chatterjee N, Pal TK (2019) Nanoliposomes of supercritical carbon dioxide extract of small cardamom seeds redresses type 2 diabetes and hypercholesterolemia. Recent Pat Biotechnol 13:284–303. https://doi.org/10.2174/1872208313666190404101336

Peters RJ, Croteau RB (2003) Alternative termination chemistries utilized by monoterpene cyclases: chimeric analysis of bornyl diphosphate, 1,8-cineole, and sabinene synthases. Arch Biochem Biophys 417:203–211

Pillai OGN, Mathulla T, George KM, Balakrishnan KV, Varghese J (1984) Studies in cardamom 2. An appraisal of the excellence of Indian cardamom (*Elettaria cardamomum* Maton.). Indian Spices 21(2):17–25

Santos FA, Rao VSN (2000) Antiinflammatory and antinociceptive effects of 1,8-cineole a terpenoid oxide present in many plant essential oils. Phytother Res 14(4):240–244. https://doi.org/10.1002/1099-1573(200006)14:4

U.S. Department of Agriculture, Agricultural Research Service (2012) USDA national nutrient database for standard reference, release 25. http://www.ars.usda.gov/ba/bhnrc/ndl

Wang H, Ma D, Yang J, Deng K, Li M, Ji X, Zhong L, Zhao H (2018) An integrative volatile terpenoid profiling and transcriptomics analysis for gene mining and functional characterization of *AvBPPS* and *AvPS* involved in the monoterpenoid biosynthesis in *Amomum villosum*. Front Plant Sci 9:846. https://doi.org/10.3389/fpls.2018.00846

Wang Q, Quan S, Xiao H (2019) Towards efficient terpenoid biosynthesis: manipulating IPP and DMAPP supply. Bioresour Bioprocess 6:6. https://doi.org/10.1186/s40643-019-0242-z

Worth H, Schacher C, Dethlefsen U (2009) Concomitant therapy with cineole (Eucalyptole) reduces exacerbations in COPD: a placebo-controlled double-blind trial. Respir Res 10(1):69. https://doi.org/10.1186/1465-9921-10-69

Yamahara J, Kashiwa H, Kishi K, Fujimura H (1989) Dermal penetration enhancement by crude drugs: *in vitro* skin permeation of prednisolone enhanced by active constituents in cardamom seed. Chem Pharm Bull 37:855–856. https://doi.org/10.1248/cpb.37.855

Zaks A, Davidovich-Rikanati R, Bar E, Inbar M, Lewinsohn E (2008) Biosynthesis of linalyl acetate and other terpenes in lemon mint (*Mentha aquatica* var. *citrata*, Lamiaceae) glandular trichomes. Israel J Plant Sci 56(3):233–244. https://doi.org/10.1560/IJPS.56.3.233

7 Transcriptomic Resources of Cardamom and Their Implications

M. Balakumaran, R. Gobu, J. Berliner, and C. Sellaperumal

Abstract

Cardamom, the 'queen of spices' has economic potential because of its domestic and international demand. There is a considerable difference in the productivity between the lead producer Guatemala and India. Also, various biotic factors, such as insects, disease-causing fungi, bacteria, virus and nematodes, and abiotic factors, such as drought, limit the production, productivity and quality of the cardamom. The real value of cardamom lies in the phytochemical compounds present in it. The metabolic pathways governing their production are unexplored. The knowledge gaps in these areas have to be addressed so as to achieve the maximum yield with desired quality produce. Transcriptomics, a tool of great importance which may serve as panacea for these shortcomings if it is combined with the other 'omics' tools such as genomics, proteomics and metabolomics. In this chapter, we have comprehended the constraints in cardamom production and the solutions brought out through transcriptomics approaches.

M. Balakumaran (✉) · C. Sellaperumal
ICAR—Indian Institute of Spices Research, Kozhikode, India
e-mail: marannkl@gmail.com; manimaran.b@icar.gov.in

R. Gobu
ICAR—Sugarcane Breeding Research Institute, Coimbatore, India

J. Berliner
ICAR—Indian Agricultural Research Institute, Regional Station, Wellington, Nilgiris, India

Keywords

Cardamom · Productivity · Biotic factors · Abiotic factors · Transcriptomics · Omics

7.1 Introduction

Cardamom (*Elettaria cardamomum* Maton) is one of the most expensive spice due to its aroma, taste and medicinal properties, aptly called the 'queen of spices'. It belongs to the Zingiberaceae family which consists of more than 1500 species (Li et al. 2021), most of them having considerable economic potential due to the presence of various high-value compounds. The fruits and seeds of cardamom are widely used in culinary preparations, aromatherapy, and also as an ingredient in traditional medicine systems (Ashokkumar et al. 2020). The essential oils present in the fruits are found to have anti-microbial effects and are also responsible for various medicinal properties associated with this crop (Alam et al. 2021; Winarsi et al. 2014; Acharya et al. 2010; Crowell and Gould 1994;

K. K. Sabu and C. Kole (eds.), *The Cardamom Genome*, Compendium of Plant Genomes,
https://doi.org/10.1007/978-3-031-71829-8_7

Asamoto et al. 2002). The essential oil present in cardamom has certain unique constituents like 1,8-cineole, α-terpinyl and linalyl acetate apart from the other common compounds present in various spice crops (Acharya et al. 2010; Lewis et al. 1966; Salzer 1975). The commercial utility of cardamom, like any other spices, is mainly because of the phytochemical compound present in it. So, the genetic improvement in cardamom has to focus not only on yield improvement but also on improving the phytochemical compounds present in it. However, the lack of robust and consolidated genomic resources in cardamom hinders its genetic improvement process, despite the huge genetic variability harboured in India. An in-depth knowledge of the genome, transcriptome and metabolome may help in unravelling the medicinal and aromatic properties associated with this economically important spice crop.

A transcriptome refers to the complete set of RNA transcripts, including messenger RNA (mRNA), ribosomal RNA (rRNA), and transfer RNA (tRNA), produced by the cells of an organism. Transcriptome generally represents the active genes in a particular tissue or at a specific developmental stage, providing a snapshot of gene expression.

Transcriptome analysis enables researchers to examine which genes are actively expressed under different conditions, like biotic and abiotic stress, or specific developmental stages. This information is crucial for understanding the molecular mechanisms underlying plant responses. Also, such data can be used to identify molecular markers associated with desirable traits, facilitating marker-assisted breeding programs. This allows for the selection of plants with desired characteristics more efficiently and accurately. Transcriptomics allows researchers to analyse entire metabolic pathways and gene networks, providing a holistic view of the interconnected processes within a plant. This information aids in optimizing metabolic pathways for improved crop performance. It can accelerate the breeding process by providing insights into the genetic basis of traits, allowing breeders to make informed decisions and streamline the development of improved crop varieties. So, transcriptome analysis plays a pivotal role in advancing crop improvement efforts by providing comprehensive insights into gene expression patterns, functional genomics, and molecular mechanisms underlying important traits (Fig. 7.1). This knowledge is instrumental in developing crops that are more resilient, productive, and better adapted to changing environmental conditions. There are only a few transcriptome studies carried out on cardamom, the compilation of which may provide comprehensive information to the cardamom crop improvement workers.

7.2 Botanical Background of Cardamom

In general, there are two types of cardamom namely green cardamom (*Elettaria cardamomum*) and black cardamom (*Amomum subulatum*). Both of these cardamom species belong to the Zingiberaceae family, which also includes other well-known spices like ginger (*Zingiber officinale*), turmeric (*Curcuma longa*), black turmeric (*Curcuma caesia*), mango ginger (*Curcuma amada*), galangal (*Alpinia galanga*), etc. Green cardamom is more common and is being widely used compared to black cardamom. Green cardamom plants grow well in warm and humid climates, typically at elevations ranging from 600 to 1500 m above sea level. They require well-drained soil and are often cultivated on hilly terrains. These plants are shade-loving herbaceous perennials, with tall shoots and lance-shaped leaves having flower-bearing panicles. The flower colour generally varies from white to pale green and may also have a yellowish tinge. The fruit capsules contain small, aromatic seeds which are harvested for use as the spice. Cardamom plants can be propagated through seeds or by the rhizomes.

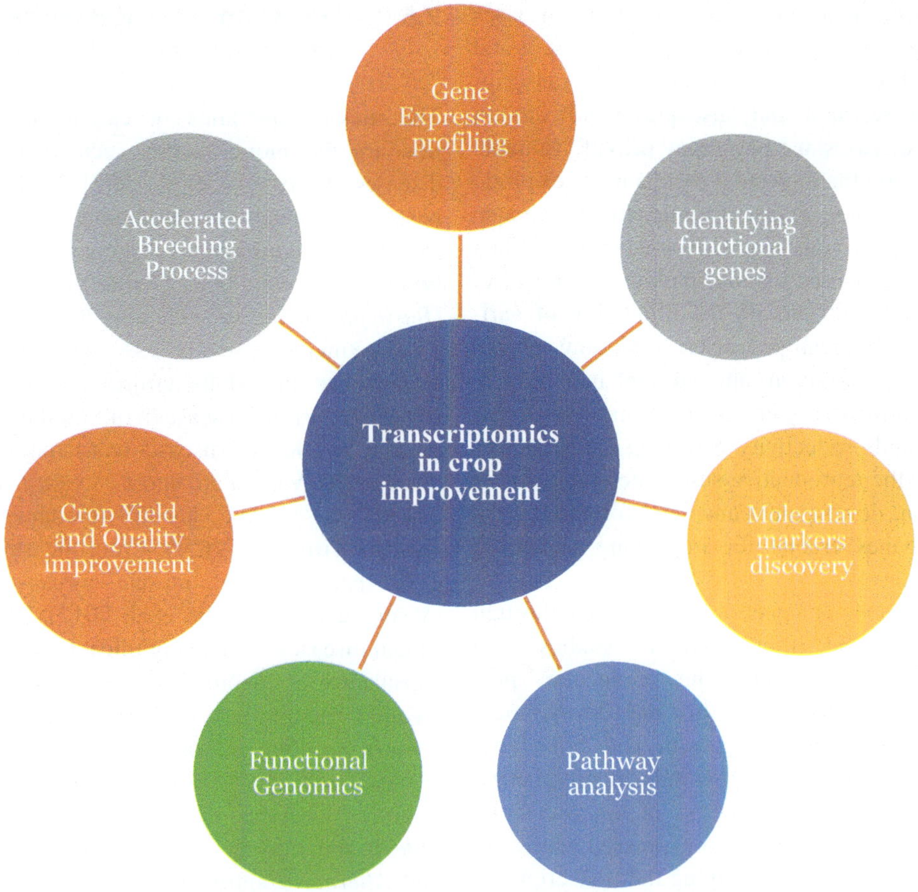

Fig. 7.1 Applications of transcriptomics in crop improvement

7.3 Key Botanical Features Relevant to Transcriptomic Studies

It is essential to focus on key botanical features that may impact gene expression and regulation of various metabolic pathways. Different tissues, such as leaves, stems, and seeds, may exhibit specific gene expression patterns. Transcriptomic studies can help identify genes that are highly expressed in particular tissues, providing insights into the molecular processes occurring in those tissues. In addition, gene expression can vary significantly during different developmental stages of the plant. Studying the transcriptome at various growth stages can uncover genes associated with processes like flowering, fruit development, and seed maturation, contributing to a comprehensive understanding of the plant's life cycle. In the rapid climate change scenario, cardamom being a shade-loving crop encounters various environmental stressors, such as changes in temperature, humidity, and light. Transcriptomic studies under different environmental conditions can reveal genes involved in stress responses, thereby enabling us to understand the plant's adaptability and also to identify tolerant lines. Cardamom is mainly valued for its essential oil, aromatic compounds and flavour. It is possible to identify genes involved in secondary metabolite pathways, such as those responsible for the synthesis of essential oils and flavour compounds. Being a vegetatively propagated crop, cardamom encounters various pests and diseases like rhizome rot, katte disease, thrips, shoot and capsule borer, etc. Understanding the transcriptome of resistant and susceptible cultivars to various diseases and pests may provide valuable insights into the molecular mechanisms of biotic stress resistance. Transcriptomic studies can also help in identifying genes involved in the plant's defence responses against pathogens and also, can provide insights on the regulation of various physiological processes, including photosynthesis, respiration, and nutrient metabolism. This information is crucial for optimizing plant growth and productivity. Studying the transcriptome of the reproductive stage, including flower initiation, development and seed formation, can reveal genes crucial for improving yield and seed quality.

Cardamom is generally propagated through rhizomes, and transcriptomic studies can provide insights into the molecular processes associated with rhizome development and the regulation of growth and branching. Transcriptomic studies also help in deciphering the gene regulatory networks of key metabolic processes by identifying co-expressed genes and transcription factors. Studying the transcriptome of different germplasm accessions can provide insights into the genetic diversity with respect to various economically important traits, aiding in crop improvement efforts. So, transcriptomic studies in cardamom have potential applications in aiding future crop improvement programmes.

7.4 Constraints in Cardamom Production and Transcriptome as a Solution

Various abiotic and biotic stresses are impediment in productivity. Among the abiotic stresses, drought, temperature extremes and nutrient loss are relevant. Among biotic stresses, viruses, fungi, bacteria, and insects are common. Different researchers tried to address some of these key issues through transcriptomics (Table 7.1).

7.5 Transcriptomic Resources Addressing Biotic Stress

The major constraints of cardamom production are the biotic factors such as pests and diseases (Praveena et al. 2021). The major pests associated with cardamom are the thrips (*Sciothrips cardamomi*), shoot and capsule borer (*Conogethes punctiferalis*) and root grub (*Basilepta fulvicorne*) (Thomas 2001).

The major diseases associated with cardamom are caused by viruses and fungi. The important viral diseases of cardamom are mosaic or 'Katte' disease (cardamom mosaic virus—CdMV), cardamom vein clearing disease or 'Kokke kandu' (cardamom vein clearing virus—CdVCV), Chlorotic streak disease (banana bract mosaic virus—BBrMV) (Venugopal 1995; Siljo et al. 2012). The major fungal diseases causing significant yield loss are capsule rot or 'azhukal' disease (*Phytophthora meadii*), rhizome rot or clump rot (*Rhizoctonia solani*), seedling rot or damping off (*Pythium vexans*), root rot (*Fusarium oxysporum*), leaf blight or 'Chenthal' (*Colletotrichum gloeosporioides*), and nursery leaf spots (*Fusarium* spp. and *Alternaria* spp.) (Thomas and Bhai 1995; Praveena et al. 2021).

7.5.1 Katte Disease

'Katte disease' in cardamom is caused by cardamom mosaic virus (CdMV) which is a species under the genus *Macluravirus* of the family Potyviridae. CdMV is a vector-borne virus transmitted by the aphid, *Pentalonia nigronervosa*. A distinct mosaic pattern created by pale green stripes of leaf is a typical symptom of this disease.

The study by Khan et al. (2020) was the first transcriptome-based approach to decipher the plant virus interaction mechanism in cardamom. A reference transcriptome of 123,338 transcripts with 43.14% GC content and 1520 bp N50 value was generated from 18,372,672 and 17,708,688 raw reads from control and infected

Table 7.1 Transcriptome resources available on cardamom

Study	Purpose	Library type	Outcome	References
Bioproject: PRJNA292483 SRA: SRR2154236 SRR2154238	To understand the domestication-related traits, gene regulation, by comparative transcriptomics of wild and cultivated genotypes of cardamom	Ion torrent proton Single end	• First comprehensive transcriptome resource on cardamom • Differences in expression pattern of key growth-related genes involved in wild and cultivated genotypes	Nadiya et al. (2016, 2018)
Bioproject: PRJNA474822 SRA: SRR7280098 SRR7280099	To understand the mechanism of cardamom and cardamom mosaic virus interaction and pathways involved in it	Illumina HiSeq 2000 Paired end	• First transcriptome-based approach to decipher the plant virus interaction mechanism in cardamom • Identified key pathways, genes and transcription factors expressed during cardamom-CdMV interactions	Khan et al. (2020)
Bioproject: PRJNA350830 SRA: SRR4457910 SRR4457911	To understand the role of microRNA in drought stress tolerance in cardamom	Ion torrent proton Single end	Deciphered the role of microRNA in drought stress response of cardamom	Anjali et al. (2017)
Bioproject: PRJNA362695 SRA: SRR5196700 SRR5196701	To understand the mechanism of disease incidence and host response to capsule rot disease infected plants	Illumina HiSeq 2000 Paired end	Comparative transcriptomic data generated for capsule rot infected and healthy cardamom plants	Mary Mathew et al. (2019)

cardamom plants, respectively. BLASTn analysis of de novo transcriptome assembly revealed that 66.50% of total transcripts, i.e. 81,966 had similarity with the genes in non-redundant (NR) database.

A total of 5317 differentially expressed genes (DEGs) that includes 2579 upregulated and 2738 downregulated genes. Among these includes key genes attributing various functions such as genes from, Resistance Gene Analog (RGA) family *RGA1*, *RGA3*; Pathogenesis Related proteins (PR) family *PR1*, *PR5*; Receptor-Like Kinase (RLK) family *NBS-LRR* a gene involved in pathogen recognition via signalling pathways; Mitogen-Activated Protein Kinase (MAPK) family *ANP1-like gene*, which involves in signal transduction; detoxification gene family *glutathione-S-transferase*; Heat Shock Protein (HSP) family *hsp70* a molecular chaperone. Although most of these genes were found upregulated, Nucleotide Binding Site—Leucine Rich Repeat (NBS-LRR) which is reported to be involved in pathogen recognition showed downregulation, which is further validated through qPCR.

The pathway analysis revealed that among the DEGs, a total of 1638 transcripts were involved in 114 metabolic pathways. Domain and family search with *InterProScan* showed that 5069 transcripts with 1219 domains and 1735 transcripts belonging to 807 protein families were found in differentially expressed transcripts. Further, a total of 2267 transcriptional factors were found out of 5317 DEGs. Among these, the transcriptional factors (TFs), *NAC*, *MYB*, *WRKY*, *bHLH*, *C2H2* were found to be most abundant. Thus, these TFs can be explored for developing disease resistance. A total of 21,735 simple sequence repeat (SSR) markers were reported in this study which can be used in genetic diversity studies. This number is far higher than the existing SSR markers (< 450) in this crop.

This study decrypted the enigma of growth retardation in the CdMV infected cardamom plants is mediated by the disturbance in auxin homeostasis based on the Gene Regulatory Network analysis in which Auxin Response Factor (ARF) has been found as a hub gene. This study also delineated the mechanism of cardamom-CdMV interaction as a cascade of events starting with virus entry recognition through RLK family genes such as LRR receptors which further activates the signalling pathways through the MAPK and ends with production of plant defence response. This is achieved by calcium-calmodulin (CaM) system a key regulator of antiviral response and *TIFY 6B* gene which mediates Jasmonic acid pathway. Whereas the HSPs serve the function of molecular chaperone and maintain protein homeostasis.

The information on defence responses mediated by different genes can be exploited for improving disease resistance in cardamom. Further, the upregulated TFs in this study such as *NAC*, *MYB*, *WRKY*, *bHLH*, *C2H2* are potentially involved in growth, development and stress tolerance of different crops (Jiang and Pan 2012; Voitsik et al. 2013). This study generated a genomic resource with putative molecular markers of key candidate genes to develop a resilient cardamom variety to withstand mosaic virus.

7.5.2 Capsule Rot

Capsule rot disease or 'Azhukal' disease was first reported in Idukki district of Kerala by Menon et al. (1972). It is one of the serious diseases in cardamom with potential yield loss to the tune of 40%. It is caused by the fungi *Phytophthora meadii* and the infected capsules either shed or become deformed (Thomas and Bhai 1995).

A comparative transcriptomics study by Mary Mathew et al. (2019) generated the raw reads of 46,931,637 and 31,682,496 from control and capsule rot-infected samples, respectively. De novo assembly of these raw reads yielded 111,495 and 91,096 transcripts from the former and latter, respectively. Their N50 values of the contigs were 3013 for control and 2729 for diseased. A reference transcriptome was generated by combining raw reads from both samples yielding a total of 162,589 contigs. The results of this study may give insights into the mechanism underlying the disease incidence and the corresponding responses in the cardamom plant.

7.6 Transcriptomic Resources Addressing Abiotic Stress

The abiotic stress factors like droughts, water logging, temperatures, salinity, nutrient deficiency and oxidative stress limit cardamom production. Among them, drought is one of the major limiting factors that significantly reduce the productivity of this crop (Anjali et al. 2019).

Anjali et al. (2019) developed a comprehensive transcriptome resource of small RNAs of cardamom to decipher the mechanism of microRNA (miRNA) mediated drought stress response. The miRNAs are small (18–25 nucleotides), highly conserved non-coding, single-stranded RNA molecules involved in the regulation of gene expression (Ying et al. 2008; MacFarlane and Murphy 2010). The comparative transcriptome study by Anjali and co-workers (2019) between control and moisture-stressed cardamom plants revealed many useful information. Raw reads of size 3,938,342 and 4,083,181 were obtained from control and stressed cardamom samples, from which, 1,251,632 and 997,432 unique reads were obtained, respectively. Most of the sequences were of size 21 nucleotide in both the samples. It reports a total of *150* conserved and 20 novel microRNAs from both the control and stressed cardamom libraries. The following miRNAs were reported to be abundantly expressed in both the libraries, miR159, miR396, miR396e, miR535, miR166b, and miR167. These miRNAs were found to target various developmental genes such as LEAFY, GRF, SPL, HD-ZIP III, ARF genes, respectively, in different plants (Reyes and Chua 2007; Boualem et al. 2008; Liu et al. 2008; Khaksefidi et al. 2015).

In total 17 miRNAs belonging to 11 families were found to be in drought stress response. Among these 17 differentially expressed miRNAs, 11 were upregulated and 6 were downregulated. A non-conserved miRNA, miR3711 was discovered for the first time to be involved in the drought stress regulation. This miRNA was upregulated in *Astragalus chrysochlorus* treated with selenium (Cakir et al. 2016). It is possible that it may be a cross-adaptation mechanism by cardamom which needs confirmation. Thus, further studies are necessary to understand the regulatory mechanism of miR3711. The miRNA target prediction revealed several target genes with water deprivation functions were identified. The miR156l and miR169c which cleave the target mRNA involved in response to water deprivation, while, the miR156a and miR530b target mRNAs which respond to water deprivation by inhibiting the translational process.

The study by Anjali et al. (2019) provided an insight into the drought-responsive miRNAs of cardamom and serves as a resource base for functional genomics. Discovery of 17 differentially expressed miRNAs delineated the various biological processes involved in drought stress that can be utilized for developing moisture-stress-tolerant cardamom varieties.

7.7 Transcriptomic Resources Addressing Crop Improvement

Nadiya et al. (2018) performed comparative transcriptomics between wild and cultivar genotypes to understand the genes and the key pathways involved in growth and development of cardamom plants. A comprehensive reference transcriptome was generated comprising 178,745 contigs. The functional annotation analysis determined by performing BLASTx search against various databases revealed that ~76% of unigenes were annotated against TAIR10 database followed by 57.5% against and 55% Uniprot/Swissprot database and KOG database, respectively. The gene ontology analysis revealed that the cardamom unigenes were assigned to 62269, 45270, and 34553 GO terms with biological process, cellular process and molecular functions, respectively. Also, the analysis of unigenes involved in regulation of transcription unfolded that 570 transcription factors representing 39 families and 155 transcription regulators representing 12 families in the cardamom transcriptome. This study provides a better understanding of the genomic resources on cardamom.

The differential expression analysis among wild and cultivated genotypes deciphered that 132 genes were differentially expressed which includes 105 upregulated and 27 downregulated genes, respectively, in cultivars. An important finding is that the wild cardamom has enhanced attributes such as defence response, drought stress tolerance, and various other aspects of plant growth regulations. In brief, the cultivated cardamom showed upregulation of genes involved in flavonoid biosynthesis pathway, while exhibited downregulation of genes encoding enzymes involved in monoterpenoid biosynthesis and resistance-related genes such as cinnamoyl CoA reductase, etc. The flavonoids are attributed to the anti-oxidant properties of cardamom and the, monoterpenes are responsible for the synthesis of various bioactive compounds. Besides, the transcription factors namely *NAC*, *NF-YA*, *NFYC*, and *WRKY* were found to be significantly up regulated in the wild genotypes in comparison with the cultivars. These regulatory elements are reported to be responsible for various abiotic and biotic stress tolerance in different crops. The comparative transcriptomics decrypted the genes involved in different metabolic pathways and created a knowledge base which can be utilized for crop improvement.

7.8 Databases Containing Transcriptome Resources of Small Cardamom

7.8.1 SmCarTDB (Small Cardamom Transcriptome Database)

The SmCarTDB houses information on comparative transcriptome of healthy and capsule rot diseased cardamom. The differentially expressed genes (DEGs), exhaustive markers, microRNAs, transcription factors (TFs) and pathways from the transcriptome of small cardamom associated with capsule rot disease, one of the most serious fungal diseases in cardamom caused by *Phytopthora meadii*. This catalogue contains 44,000 genic regions putative SSR markers.

7.8.2 SCMVTDb: Small Cardamom Mosaic Virus Transcriptome Database

SCMVTDb is a database that contains information on comparative transcriptome of small or green cardamom infected mosaic virus. It is comprehensive and encompasses differentially expressed genes, SSRs, transcription factors, pathways, domain and families. The mosaic or 'katte' or 'marble disease' is the most destructive disease in small cardamom and is caused by Cardamom mosaic virus (CdMV). The database contains a total of 123,338 transcripts, 5317 DEGs, 24,000 genic region putative markers, 2267 transcription factors, 147,442 SNPs and 154,217 Indels from healthy and virus-infected samples.

7.9 Conclusions and Future Prospects

Cardamom a crop with commercial importance and growing demand has many untapped potentials. The underexplored genetic potential of the crop may address the production constraints and quality improvement. Transcriptomics serves as a sovereign remedy to these limitations. Key highlights achieved so far through this approach include the delineation of mechanisms underlying cardamom-CdMV interaction, generation of data on 'capsule rot' infection process, role of microRNAs in drought stress regulation, and traits associated with growth and development in cultivar compared to wild genotype, etc. This functional genomics tool opened up many new avenues that can be explored for the crop improvement. However, still many questions remain unanswered such as the mechanism of genes governing the metabolic pathways producing the phytochemicals which contribute to the aroma and medicinal properties of the cardamom; the search for resistance and the underlying mechanisms against various biotic stress factors; the quest for developing a resilient cultivar that can withstand the abiotic stress factors. A combinatorial approach including the other functional genomics tools such as metabolomics, and proteomics may serve as a 'golden fleece' for addressing these needs.

References

Acharya A, Das I, Singh S, Saha T (2010) Chemopreventive properties of indole-3-carbinol, diindolylmethane and other constituents of cardamom against carcinogenesis. Recent Pat Food Nutr Agric 2(2):166–177. https://doi.org/10.2174/2212798411002020166

Alam A, Rehman NU, Ansari MN, Palla AH (2021) Effects of essential oils of *Elettaria cardamomum* grown in India and Guatemala on gram-negative bacteria and gastrointestinal disorders. Molecules 26(9):2546. https://doi.org/10.3390/molecules26092546

Anjali N, Nadiya F, Jinu T, Sabu KK (2017) Discovery of MicroRNAs in cardamom (Elettaria cardamomum Maton) under drought stress. Dataset Pap Sci. https://doi.org/10.1155/2017/9507485

Anjali N, Nadiya F, Thomas J, Sabu KK (2019) Identification and characterization of drought responsive microRNAs and their target genes in cardamom (*Elettaria cardamomum* Maton). Plant Growth Regul 87:201–216. https://doi.org/10.1007/s10725-018-0462-9

Asamoto M, Ota T, Toriyama-Baba H, Hokaiwado N, Naito A, Tsuda H (2002) Mammary carcinomas induced in human c-Ha-ras proto-oncogene transgenic rats are estrogen-independent, but responsive to d-limonene treatment. Jpn J Cancer Res 93(1):32–35. https://doi.org/10.1111/j.1349-7006.2002.tb01197.x

Ashokkumar K, Murugan M, Dhanya MK, Warkentin TD (2020) Botany, traditional uses, phytochemistry and biological activities of cardamom [*Elettaria cardamomum* (L.) Maton]—a critical review. J Ethnopharmacol 246:112244. https://doi.org/10.1016/j.jep.2019.112244

Boualem A, Laporte P, Jovanovic M, Laffont C, Plet J, Combier JP, Niebel A, Crespi M, Frugier F (2008) MicroRNA166 controls root and nodule development in *Medicago truncatula*. Plant J Cell Mol Biol 54(5):876–887. https://doi.org/10.1111/j.1365-313X.2008.03448.x

Cakir O, Candar-Cakir B, Zhang B (2016) Small RNA and degradome sequencing reveal important microRNA function in *Astragalus chrysochlorus* response to selenium stimuli. Plant Biotechnol J 14(2):543–556. https://doi.org/10.1111/pbi.12397

Crowell PL, Gould MN (1994) Chemoprevention and therapy of cancer by d-limonene. Crit Rev Oncog 5(1):1–22. https://doi.org/10.1615/critrevoncog.v5.i1.10

Jiang L, Pan LJ (2012) Identification and expression of C2H2 transcription factor genes in *Carica papaya* under abiotic and biotic stresses. Mol Biol Rep 39:7105–7115. https://doi.org/10.1007/s11033-012-1542-y

Khaksefidi RE, Mirlohi S, Khalaji F, Fakhari Z, Shiran B, Fallahi H, Rafiei F, Budak H, Ebrahimie E (2015) Differential expression of seven conserved microRNAs in response to abiotic stress and their regulatory network in *Helianthus annuus*. Front Plant Sci 6:741. https://doi.org/10.3389/fpls.2015.00741

Khan A, Johnson George K, Jasrotia RS, Aravind S, Angadi UB, Iquebal MA, Manju KP, Jaiswal S, Umadevi P, Rai A, Kumar D (2020) Plant virus interaction mechanism and associated pathways in mosaic disease of small cardamom (*Elettaria cardamomum* Maton) by RNA-Seq approach. Genomics 112(2):2041–2051. https://doi.org/10.1016/j.ygeno.2019.11.017

Lewis YS, Nambudiri ES, Philip T (1966) Composition of cardamom oils. Perfum Essent Oil Rec 57:623–628. http://ir.cftri.res.in/id/eprint/2779

Li DM, Li J, Wang DR, Xu YC, Zhu GF (2021) Molecular evolution of chloroplast genomes in subfamily Zingiberoideae (Zingiberaceae). BMC Plant Biol 21:1–24. https://doi.org/10.1186/s12870-021-03315-9

Liu HH, Tian X, Li YJ, Wu CA, Zheng CC (2008) Microarray-based analysis of stress-regulated microRNAs in *Arabidopsis thaliana*. RNA (New York, N.Y.) 14(5):836–843. https://doi.org/10.1261/rna.895308

Macfarlane LA, Murphy PR (2010) MicroRNA: biogenesis, function and role in cancer. Curr Genom 11(7):537–561

Mary Mathew K, Reshma R, Geethu M, Rithin V, Sabu KK, Nadiya F, Noushad MA, Dharan SS, Rao YS, Remashree AB (2019) Data on small cardamom transcriptome associated with capsule rot disease. Data Brief 27:104625. https://doi.org/10.1016/j.dib.2019.104625

Menon MR, Sajoo BV, Ramakrishnan CK, Remadevi L (1972) A new Phytophthora disease of cardamom. Curr Sci 41:231

Nadiya F, Anjali N, Thomas J, Gangaprasad A, Sabu KK (2016) Transcriptome profiling of Elettaria cardamomum (L.) Maton (small cardamom). Genom Data 11:102–103. https://doi.org/10.1016/j.

gdata.2016.12.013. PMID: 28070488; PMCID: PMC5219599

Nadiya F, Anjali N, Thomas J, Gangaprasad A, Sabu KK (2018) Genome-wide differential expression profiling in wild and cultivar genotypes of cardamom reveals regulation of key pathways in plant growth and development. Agri Gene 8:18–27. https://doi.org/10.1016/j.aggene.2018.03.002

Praveena R, Biju CN, Sujatha AM (2021) Deciphering primary incitant of rot complex of small cardamom plant by sequential inoculation of pathogens. Indian Phytopathol 74:133–143. https://doi.org/10.1007/s42360-020-00309-9

Reyes JL, Chua N (2007) ABA induction of miR159 controls transcript levels of two MYB factors during Arabidopsis seed germination. Plant J 49:592–606. https://doi.org/10.1111/j.1365-313X.2006.02980.x

Salzer U (1975) Analytical evaluation of seasoning extracts (oleoresins) and essential oils from seasoning. II. Int Flavours Foods Addit 6(4):206–210

Siljo A, Bhat AI, Biju CN, Venugopal MN (2012) Occurrence of banana bract mosaic virus on cardamom. Phytoparasitica 40:77–85

Thomas J (2001) Biological control of insect pests of small cardamom. In: Upadhyay RK, Mukerji KG, Chamola BP (eds) Biocontrol potential and its exploitation in sustainable agriculture. Springer, Boston, MA. https://doi.org/10.1007/978-1-4615-1377-3_23

Thomas J, Bhai RS (1995) Fungal and bacterial diseases of cardamom (*Elittaria cardamomum* Maton) and their management. J Spices Aromat Crops 4(1):24–31

Venugopal MN (1995) Viral diseases of cardamom (*Elettaria cardamomum* Maton) and their management. J Spices Aromat Crops 4:32–39

Voitsik AM, Muench S, Deising HB, Voll LM (2013) Two recently duplicated maize NAC transcription factor paralogs are induced in response to *Colletotrichum graminicola* infection. BMC Plant Biol 13:85. https://doi.org/10.1186/1471-2229-13-85

Winarsi H, Sasongko ND, Purwanto A, Nuraeni I (2014) Effect of cardamom leaves extract as antidiabetic, weight lost and hypocholesterolemic to alloxan-induced *Sprague dawley* diabetic rats. Int Food Res J 21:2253–2261

Ying SY, Chang DC, Lin SL (2008) The microRNA (miRNA): overview of the RNA genes that modulate gene function. Mol Biotechnol 38(3):257–268. https://doi.org/10.1007/s12033-007-9013-8

Molecular Characterization and Phylogenetic Relationship of Cardamom from Colombia and Costa Rica

8

Ana Cristina Tamayo Domínguez

Abstract

Cardamom in Colombia had a long process of conventional improvement, where material selection tests were carried out with the best harvest results, both in quantity and quality, obtaining an improved cardamom material, which was characterized by AFLP molecular markers in the year 2003. This improvement program began in the field in plantations in Colombia, in the department of Antioquia 30 years ago, carrying out a continuous process of selection of cardamom materials that showed a high productive potential, the plantations were separated into various lots within the same farm, which had an extension close to 1300 ha, the plants were crossed with each other and conserved for their subsequent evaluation of production and quality, and through repeated crossings and monitoring of the plantations, a clonal selection process was established for the crop. In this program, promising materials in productivity, color and capsule size are obtained, which were analyzed by the same author of this writing in 2003, using AFLP molecular markers, to determine polymorphisms between them, where she finds the majority of improved materials; Said breeding program and some data presented here are part of her research published 20 years ago, whose reference is presented in this chapter as Tamayo (Caracterización molecular de materiales seleccionados de cardamom—*Elettaria cardamomum* L. (Matón)—utilizando marcadores AFLP. Universidad Nacional de Colombia, Facultad de Ciencias Seccional Medellín, Trabajo de grado de Maestría en Biotecnología, 2003). In this study, improved, common, wild materials and a sample from Costa Rica were included, where it had been determined to be the ancestor of the Colombian material. After 20 years, some samples from that time and other current ones are selected, which were subjected to sequencing of the ITS2 region, and three more accessions from Costa Rica were included, located in two specific sites in that country. The phylogenetic tree obtained by correlating the sequences of the ITS2 region of the Colombian and Costa Rican materials, in 2003 and 2023, is analyzed to visualize the relationships between ancestral lines and current lines of cardamom. The genetic sequences of the evaluated materials are included as a source for further research.

A. C. T. Domínguez (✉)
EARTH University, Guácimo, Costa Rica
e-mail: atamayo@earth.ac.cr

K. K. Sabu and C. Kole (eds.), *The Cardamom Genome*, Compendium of Plant Genomes,
https://doi.org/10.1007/978-3-031-71829-8_8

Keywords

Cardamom · Molecular characterization · Phylogenetic · ITS2 región

8.1 Overview

Colombia is a country in South America with land suitable for cardamom farming. In 2003, the lands planted with cardamom were located in the southwestern region of the department of Antioquia, in the municipality of Jericó.

Costa Rica is a country in Central America and, in 2023, the lands planted with cardamom are located in the canton of Guatuso, province of Alajuela.

According to Lüttmann (1985), cardamom was introduced to Central America in 1914, from where it was exported to Colombia. The variety from which the seeds originated is unknown, but in the samples from both countries—Colombia and Costa Rica—evaluated in 2003 and again in 2023, the plants have thin leaves, long stems, tall flower spikes and fruit of variable size, some large and long and others small and round.

The cardamom flower (Fig. 8.1) requires the presence of external pollinators (Lang 1982). The longer the flowers are exposed to the pollinators, the higher the percentage of fruiting.

From 1993 to 2001, the most economically important pest observed to have affected the plantations in the department of Antioquia was the weevil (*Cholus* spp.) and, collaterally, some defoliators. As for diseases, the most limiting is basal rot caused by the association of *Fusarium* spp. and *Pythium* spp. (Tamayo 2003).

Starting in 2001, cardamom thrips (*Sciothrips cardamomi* (Rank)) emerged as a severe pest causing 60–90% of losses in cardamom crops (Dharmadasa et al. 2008). According to Jacob et al. (2015), it is the most destructive and persistent pest. The damage caused to capsules by thrips is seen in Fig. 8.2. Thuruthumkara et al. (2020) report the existence of pest-resistant varieties.

Once harvested, cardamom undergoes a drying process and is classified by size and color, with preference given to large, green capsules, followed by pale-colored capsules and, finally, capsules with defects, such as those that are yellow in color or open. The loose seed is considered the final product that is sold on the market (Tamayo 2003). Grading determines the percentages of superior and inferior quality for the market, where capsules with scars or bites caused by thrips or weevils lose market value.

In the plantations of Antioquia, Colombia, harvesting is staggered on a monthly or bimonthly basis, which ensures superior quality due to the production of homogeneously mature capsules.

Fig. 8.1 The cardamom flower. Photograph taken by the author for this research in plantations in Guatuso, Costa Rica

Fig. 8.2 The damage caused to capsules by thrips. Note the scar along each capsule. Photograph taken by the author for this research in plantations in Guatuso, Costa Rica

In Costa Rica, specifically on the campus of EARTH University, where temperatures rise above 28 °C (EARTH University 2023), the plants labeled CR 17 exhibit fruiting problems and flowers that die before pollination. In the canton of Guatuso, there is sufficient pollination to produce spikes with capsules (Annex 1); however, the plants are affected by thrips, which leave external scars along the capsules (Fig. 8.2 and Annex 1). Moreover, there are spikes that have not been harvested, causing the capsule to turn yellowish in color (Fig. 8.3). The damage caused by thrips and the delayed harvesting cause considerable losses for farmers as they result in a final product of poor quality.

In 2023, cardamom plant material was found in two different areas of Costa Rica. Some of this material had been present since 2003 and was propagated asexually, while others came from plantations that were unrelated to each other. For this reason, the objective is to compare the cardamom materials present in Costa Rica with samples that have undergone a genetic improvement process.

Fig. 8.3 Cardamom spikes that have not been harvested, causing the capsule to turn yellowish in color. Photograph taken by the author for this research in plantations in Guatuso, Costa Rica

8.2 Genetic Plant Breeding Program for Cardamom Crops in Colombia

The improvement program described below was carried out by the author in search of improved cardamom lines, which once analyzed, needed to be compared to the current lines of the same crop. The following is a synopsis of the study conducted by Tamayo and presented in 2003, which supports the use of the plant material in the current study.

The aforementioned research states that in Colombia cardamom farming began in the 1990s as an alternative to diversity the country's coffee-growing regions, especially in the department of Antioquia, by a private company, which, at that time, was a big name in the agriculture sector in that area, with a plantation of 1300 ha and a source of employment for over 300 workers and other producers in the region. The plant material used to initiate farming in the department was imported from Costa Rica.

In these plantations, there are plants with tall stems, long flower spikes and oblong, triangular fruit, which are being monitored to create a local improvement program that will work with select cardamom plant material from populations obtained from natural crosses of introduced clones, with the aim to improve characteristics for high productivity and capsule quality, evaluating both desirable and undesirable parameters and agronomic aspects.

Cardamom is propagated both sexually and asexually. Sexual propagation by seed and asexual propagation by in vitro multiplication and by rhizomes, with the advantage of similarity with the source material.

In the department of Antioquia, Colombia, during the development of this research, an average productivity of 1.2 dry kilos per plant per year was obtained, for a total of 1538 dry kilos/ha/year. The plants selected for the genetic improvement program exceeded 5.5 dry kilos/plant/year, equal to 7050 dry kilos/ha/year with planting distances of 3 m by 3 m. The cardamom improvement program includes plant physiology studies, agronomic requirements, morphological characterization, the establishment of populations with the mixture of promising clones and the molecular characterization of the selected materials.

The materials selected for the plant breeding program exhibit desirable characteristics for the international market in terms of capsule color and size. The plants obtained in that generation again undergo a selection process, taking into account quality and quantity parameters as well as agronomic and phytosanitary aspects. A program of recurrent selection and, later, clonal selection is established and, once the fruit has developed, the material is selected in terms of quality, productivity and agronomic aspects. Plantings are made in new plots and, once production begins, the process begins again. This methodology requires years of rigorous evaluation and follow-up to avoid erroneous data.

Figure 8.4 shows the morphological characteristics evaluated in the 2003 program for the classification of cardamom according to capsule size and color, which were the basis for selecting genotypes from Colombia.

The percentages of superior and inferior cardamom quality, which in any given crop are normally 60% and 40%, respectively, and are determined by capsule color and size, are taken in the plants selected for the cardamom improvement program in the department of Antioquia, in which above-average percentages of superior quality are observed, some even higher than 90% (Table 8.1). This results in plants with superior quality and quantity of capsules, which makes investment in the crop more profitable for farmers in the area and for the country as a whole, as they will have a competitive advantage in the international market with this variety (Tamayo 2003). Molecular characterization tests are therefore needed to determine if there is a genotypic difference in the plant material.

The molecular test established for evaluating cardamom plants in 2001, based on DNA extraction from the selected plants, is the use of AFLP molecular markers, a technique reported by Vos et al. (1995) as novel with a high capacity to obtain DNA patterns of any origin and complexity. One advantage of the AFLP technique is the number of polymorphisms obtained in the amplification with the combination of primers.

In characterization with the use of AFLP markers, Tamayo (2003) works with 15 accessions of *Elettaria cardamomum* (L.) Maton from Colombia and Costa Rica, including nine promising plant materials categorized as IMPROVED from Colombia; two from Colombia and one from Costa Rica categorized as COMMON; two intermediates between common and improved categorized as INTERMEDIATE; and one wild plant material from Colombia. A brief description of each of the accessions and productivity parameters is presented in Table 8.1.

Fig. 8.4 Morphological characteristics evaluated in the 2003 program for the classification of cardamom from Colombia according to capsule size and color. **a–c** According to the size: **a** small type grain; **b** medium type grain and **c** large type grain, and **d–f** according to the color: **d** pale green grain; **e** green grain and **f** intense green grain. Photographs taken by Tamayo (2003)

Table 8.1 Percentages of superior and inferior quality obtained in the plants of the cardamom breeding program in the department of Antioquia, Colombia

Code	Category	Dry kg/pl./year	% higher quality	% lower quality	Origin
CM 1	Improved	5.6	71.52	28.45	Colombia
CM 2	Improved	6.1	81.78	18.2	Colombia
CM 3	Improved	4.3	95.79	4.2	Colombia
CM 4	Improved	8.1	82.31	17.67	Colombia
CM 5	Improved	6.1	72.39	27.59	Colombia
CM 6	Improved	6.9	91.46	8.5	Colombia
CM 7	Improved	5.0	85.46	14.51	Colombia
CM 8	Improved	> 4	–	–	Colombia
CM 9	Improved	> 4	–	–	Colombia
CM 10	Intermediate	> 1.2 < 4	60.5	39.5	Colombia
CM 11	Intermediate	> 1.2 < 4	60.5	39.5	Colombia
CM 12	Common 1	1.2	60.5	39.5	Colombia
CM 13	Common 2	1.2	60.5	39.5	Colombia
CM 14	Wild type	–	–	–	Colombia
CR 15	Common 3	–	–	–	Costa Rica

The cardamom samples for the study are collected from the plantations in Colombia, while the plant material from Costa Rica is taken as a single sample, found in the plantations eliminated in the country in 2003, and taken to EARTH University for conservation. Annex 2 includes the morphological and capsule quality characteristics of the representative accessions, with photographs taken by Tamayo (2003).

8.2.1 Selection of Promising Plant Material

The material evaluation process described by Tamayo (2003) was carried out and seven accessions with excellent characteristics were selected, labeled CM 1–CM 7; CM 8 and CM 9 were also included (Annex 2) as they have important characteristics for the improvement program; CM 9 is the result of in vitro propagation conducted by a commercial laboratory in Costa Rica in 1995 at the request of the Colombian developer, using plants from its plantations in the department of Antioquia. While the percentages of superior and inferior quality were not evaluated, these last two selected materials show important differences in phenotype, making them eligible for the study. Annex 3 shows a typical plant from the Colombian improvement program.

The materials from the previous study, labeled INTERMEDIATES CM 10 and CM 11 (Table 8.1 and Annex 2), are considered acceptable for production and show an increase in quality according to the selection parameters. COMMON CM 12 and CM 13 (Table 8.1 and Annex 2) exhibit a production close to the average for the entire plantation, but do not stand out in the aspects evaluated, and the material from Costa Rica—labeled CR 15—is included in the COMMON category seeking correlation with the other material. Subsequently, it is observed that CR 15 presents spikes with few capsules that are small and round in shape and pale green in color.

8.2.2 AFLP Molecular Marker Analysis

DNA extraction from the samples analyzed in 2003 was carried out using the methodology proposed by Harris (1996) and modified by Phillips-Mora (1995), based on the use of CTAB buffer. The methodology proposed by Astorga and Quirós (2003), cited by Tamayo (2003), is used for the digestion, ligation, preamplification and amplification of the 15 accessions using AFLP markers and EcoRI/MseI primer combinations for the evaluation of polymorphisms. Figure 8.5 shows a polyacrylamide gel electrophoresis for the preliminary evaluation of polymorphisms with 16 EcoRI/MseI primer combinations in accessions CM 7, CM 11 and CM 13. A total of 82 polymorphic AFLP markers were generated in all combinations.

In that study, the electrophoretic patterns of each gel were evaluated using a binary system designating the presence and absence of bands as 1 and 0, respectively. The data were analyzed in MVSP-MultiVariate Statistical Package, the Jaccard similarity coefficient was determined and the grouping analysis (Fig. 8.6) was performed based on the similarity matrix using the Unweighted Pair Group Method of Arithmetic Means (UPGMA) method (Bhat et al. 1995).

The dendrogram obtained in this study shows the formation of small groups. Group I is comprised of accessions CM 5, CM 7, CM 3, CM 6, CM 2, CM 8 and CM 1, corresponding to the IMPROVED category—promising materials from the developer's improvement program, but from which CM 4 and CM 9, also characterized as promising according to morphology and agronomic behavior, are excluded. Group II comprises CM 4, CM 10, CM 11 and CM 12, characterized morphologically and agronomically as INTERMEDIATE and COMMON. The material from Costa Rica (CR 15) is related to the material from Colombia and closest to CM 13, also considered COMMON. CM 9 together with commons CM 13 and CR 15 each formed different groups that were poorly related to the improved material (Tamayo 2003).

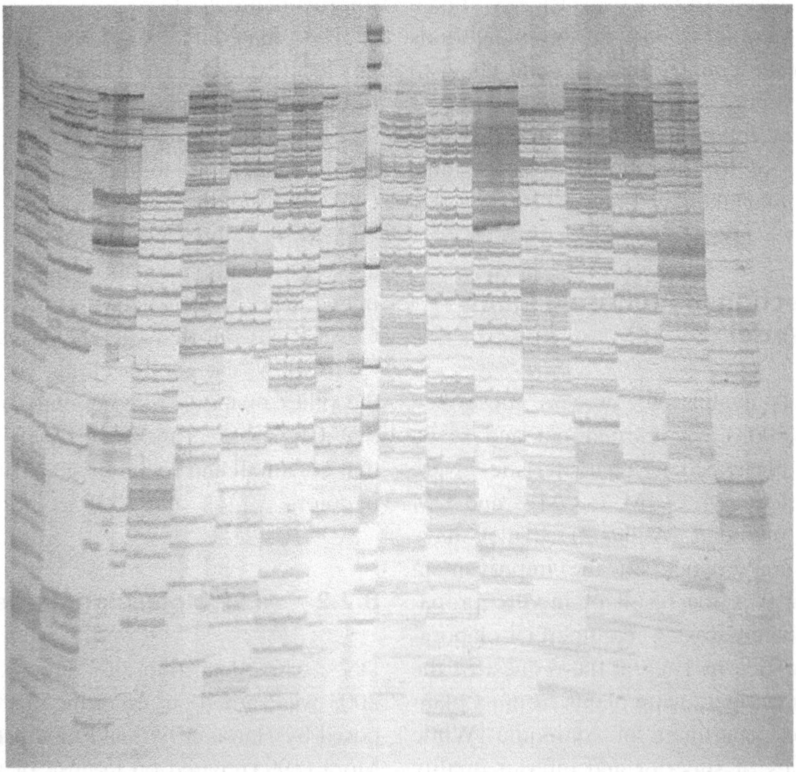

Fig. 8.5 Electrophoresis for the evaluation of polymorphisms with 16 EcoRI/MseI primer combinations in accessions CM 7, CM 11 and CM 13 (Tamayo 2003)

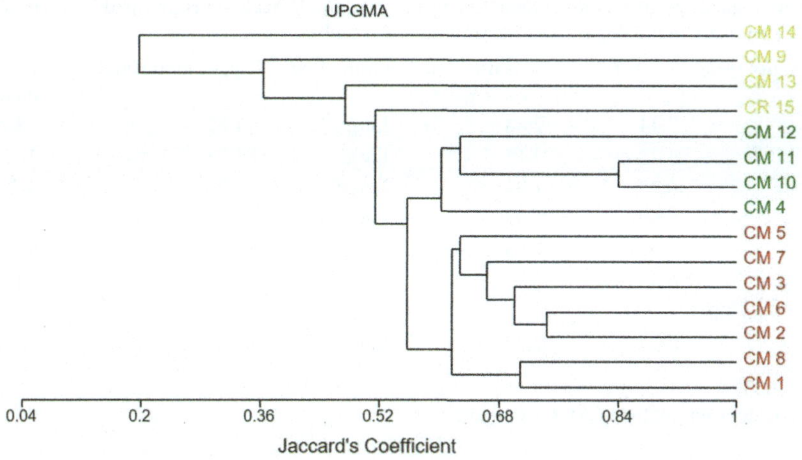

Fig. 8.6 Dendrogram based on Jaccard's similarity coefficient between 15 cardamom accessions. Cophenetic evaluation coefficient = 0.962 (Tamayo 2003)

8.2.3 Identification of Genetically Improved Lines

The field data for the plant materials show differences in productivity and quality. Accessions CM 1, CM 2, CM 3, CM 5, CM 6 and CM 7 were outstanding in all parameters, with high production volumes expressed in dry kilos/plant/year, a below average open capsule percentage and a below average spotted capsule percentage, with the exception of CM 2, which exceeds the average. The capsule type is long, which increases the percentage of superior quality. Even CM 3 exhibited 10.51% of "Extra" quality—considered the best quality—even though it has lower production in terms of dry kilos/plant/year among the improved varieties, it is a productive unit that improves the producer's income (Tamayo 2003).

Accession CM 9 is the most genetically distant from the other Group I samples, from which it was considered to have originated. At the time, it was assumed that it had undergone a different improvement process, which would explain the phenotypic difference in size and its use was suggested to increase genetic variability due to its productive characteristics; however, other tests at the genetic level are required to determine the origin (Tamayo 2003).

Accessions CM 13 and CR 1—the first from Colombia and the second from Costa Rica—both categorized as COMMON—are genetically close. At the time, in the research developed by Tamayo in 2003, it was considered that these were related given that the material from Costa Rica is the origin of the seed used in the plantations in Colombia, which is why they are included in the 2023 study.

Accessions CM 1, CM 2, CM 3, CM 5, CM 6, CM 7 and CM 8 studied in 2003 correspond to improved lines in the aforementioned plantations and present genetic variability from the other samples. Furthermore, according to the clonal selection process, they form a significant improved genetic line for producers given their high levels of production and quality (Table 8.2) and are candidates to continue in future improvement and research programs (Tamayo 2003).

8.2.4 Current Situation of the Plantations in Colombia

With this genetically improved material, the developer began an extension and promotion program to expand plantations in the southwestern region of Antioquia, Colombia, bringing

Table 8.2 Morphological and agronomic characteristics of cardamom accessions grouped by their genetic similarity according to AFLP markers in the department of Antioquia, Colombia

Code	Dry kg/pl./year	% open gr.	% stained gr.	Grain type	Upper grade quality	Lower grade quality
CM 1	5.6	7.44	2.48	Largo	71.52	28.45
CM 2	6.1	6.23	3.11	Largo	81.78	18.2
CM 3	4.3	6.3	1.96	Largo	95.79	4.2
CM 5	6.1	5.69	1.97	Largo	72.39	27.59
CM 6	6.9	5.05	1.88	Largo	91.46	8.5
CM 7	5.0	6.93	2.82	Largo	85.46	14.51
CM 8	More than 4.0	–	–	–	–	–
Average[a]	5.66	6.27	2.37	Largo	83.07	16.91

Source Tamayo (2003)
[a] The average value does not include CM 8 accession

onboard a few other producers to farm cardamom. However, in the years that followed, despite having such good quality, due to economic issues related to exports and the fact that the crop was not widely consumed locally, the developer decided to convert the majority of its lands into other economic activities, such as tourism and livestock. As a result, in the area, cultivation decreased to the point that only a few plants remain, which are maintained along with other subsistence crop plantations.

The subsequent study is intended to analyze the accessions from Colombia and Costa Rica, both those that were part of the cardamom improvement program in Colombia in 2003 and those collected in 2023.

8.3 Cardamom Cultivation in Costa Rica

In Costa Rica, the current situation and history of cardamom was investigated in 2017. According to Tapia Delgado (2017), the crop took hold as an alternative for producers in 1978 and grew to cover over 2000 ha; however, due to high production costs and broken trade relations with buyer countries, production is only sold through intermediaries, resulting in the virtual disappearance of the crop in the country. In his study, he located cardamom plants in different parts of the country, but there were no large plantations. The plants found in 2017 were mostly kept to ensure conservation of the species, not for large-scale commercialization.

In 2017, Tapia located plants at CATIE (Tropical Agricultural Research and Higher Education Center) in the province of Cartago, in the botanical garden, in a plant nursery in the province of Heredia as an exotic crop, and at EARTH University located in Las Mercedes, Guacimo, in the province of Limon, where the plants were kept for conservation purposes (Tapia Delgado 2017).

In 2023, there are no large-scale cardamom plantations in Costa Rica. The plants included in this study (Fig. 8.7) are found in the canton of Guatuso, province of Alajuela, where two producers, with a farm of no more than 5 ha, have been planting the crop since 2020. Producers mentioned that it does not have the expected result and is not marketed. In August of 2023, they reported their intention to eliminate the plants in order to plant crops of greater interest in the area and that are easier to sell. Some of their plants were relocated to EARTH University to conserve the germplasm.

For the study at hand, samples of these plants were taken and a record was made of the length and width of the capsules. Total production was not available. It is observed that the plants have a green, elongated capsule, but have a high incidence of thrip attacks—almost all samples collected have damage caused by this insect (Annex 1). With the data on length and width, by means of descriptive statistics, it is observed

Fig. 8.7 Type of plant included in this study. Photograph taken by the author for this research in plantations in Guatuso, Costa Rica

that the sample labeled number 1 exhibits significant differences (Annex 4), in addition to being phenotypically of green, elongated capsules, so this sample, cataloged as CR 18, is included for further sequencing studies.

The plants located at EARTH University in Costa Rica in 2023 belong to the Organic Production Academic Unit (UAPO) and the germplasm bank for Musa and related species, maintained for research and conservation of the species. These plants are identified as CR 16 and CR 17, respectively, and were recently relocated and have not yet started production (Fig. 8.8).

8.4 Characterization of Cardamom Plant Material by Sequencing the ITS2 Region

For molecular characterization using the ITS2 region, samples from the study conducted in Colombia and others from Costa Rica were included to determine their relationship to each other, which is expected with CR 15 due to its origin and phenotype.

The samples from Costa Rica, namely CR 16 and CR 17, were analyzed in 2022. Also included were samples CM 12 and CR 15, sequenced using the same protocol, the results of which are presented in Annex 5. Subsequently, in 2023, sequencing of the ITS2 region was performed on some samples identified from the research conducted in Colombia in the department of Antioquia, and on CR 18 obtained in Costa Rica, which according to

Fig. 8.8 Plants located in the UAPO of EARTH University. Photograph taken by the author for this research in Costa Rica

the statistical analysis was the line of greatest interest (Annex 4), all with the aim of analyzing the relationship between the materials. Both sequencing processes were conducted by the National Center for High Technology (CeNAT) in Costa Rica.

The methodology applied in 2022 to sequence the ITS2 region consisted of DNA extraction using the DNeasy® Plant Mini Kit from QIAGEN and quantification by means of the NanodropTM 2000 spectrophotometer. Following DNA extraction, polymerase chain reaction (PCR) was performed using primers for amplification of the ITS region. The PCR products obtained then underwent the sequencing protocol, which consists of cleaning the PCR products with alkaline phosphatase and exonuclease enzymes, then performing the sequencing reaction using the BigDye Terminator v3.1 kit (Applied Biosystems). Next, the products were cleaned with the BigDye X-Terminator kit (Applied Biosystems) and, finally, capillary electrophoresis was performed in a 3500 genetic analyzer (Applied Biosystems).

The sequences obtained of the cardamom samples selected were then analyzed with Sequencing Analysis Software, as well as manually using BioEdit and Finch TV software. The sequences are compared against the National Center for Biotechnology Information (NCBI) database, using the Basic Local Alignment Search Tool (BLAST). The results of sequence alignment are presented in Annex 5.

For the samples analyzed in 2023, DNA isolation was performed using the Thermo Scientific GeneJET Plant Genomic DNA Purification Kit. Subsequently, polymerase chain reaction (PCR) was performed using primers for amplification of the internal transcribed spacer region of the plant genome.

Figure 8.9 shows the result of the agarose gel for the PCR products obtained. Based on this, some of the samples were selected for the sequencing protocol, which consists of cleaning the PCR products with alkaline phosphatase and exonuclease enzymes, followed by a sequencing reaction with the BigDye Terminator v3.1 kit (Applied Biosystems). The products were then cleaned with the BigDye X-Terminator kit (Applied Biosystems) and, finally, capillary electrophoresis is performed in a SeqStudio Genetic Analyzer (A34274, Applied Biosystems).

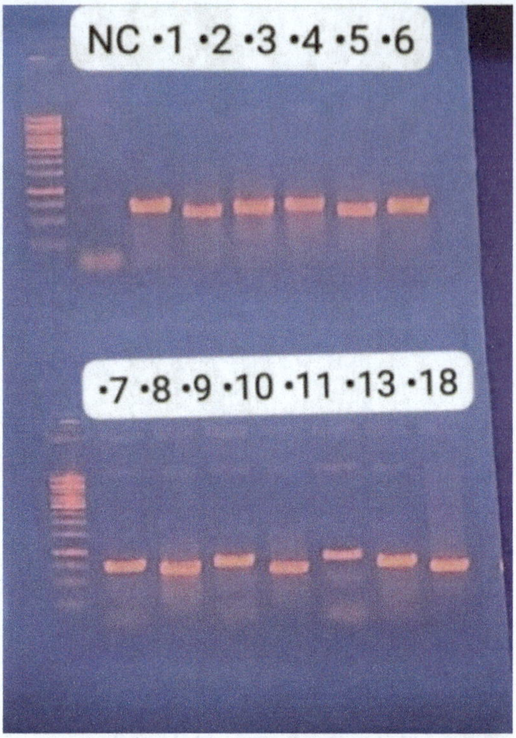

Fig. 8.9 Agarose gel, showing the integrity of PCR products from cardamom plant samples. NC: nuclease-free water negative control. Photograph taken at the CeNAT laboratory for this research, 2023

The sequences obtained were analyzed with Sequencing Analysis Software, as well as manually with the BioEdit software and automatically with R-Studio's "sangeranalyseR" package. The sequences were compared with the database of the National Center for Biotechnology Information (NCBI) using Basic Local Alignment Search Tool (BLAST).

Using both methodologies—the one applied in 2022 and the one used in 2023—sequences of the ITS2 region were obtained for the samples of cardamom selected from Colombia and Costa Rica (Annex 6).

8.5 Phylogenetic Relationship of Cardamom from Colombia and Costa Rica

The samples included for sequencing of the ITS2 region are the following:

- Sample CM 8, which is representative of the IMPROVED plant materials from Colombia in 2003.
- Sample CM 9, which was included in the Colombian improvement study and came from an in vitro propagation program in Costa Rica in 2003.
- Samples CM 10, CM 11, CM 12 and CM 13, which are representative of INTERMEDIATE and COMMON plant materials from the genetic improvement program in Colombia, to evaluate if they are grouped with any of the current samples from Costa Rica given their phenotypes and origin.
- Sample CR 15, as the original sample taken from the plantations in Costa Rica in 2003 and of which plant material is kept in cryopreservation.
- Samples CR 16 and CR 17 obtained from the plants found at EARTH University in 2023. The aim is to determine if they correlate to any of the samples given their origin.
- Sample CR 18 from the Guatuso region in Costa Rica.

The sequences obtained of the ITS2 region of the cardamom samples were introduced into the MEGA software, where grouping is determined using the Maximum Parsimony method to obtain a phylogenetic tree showing the relationship between the different samples (Fig. 8.10).

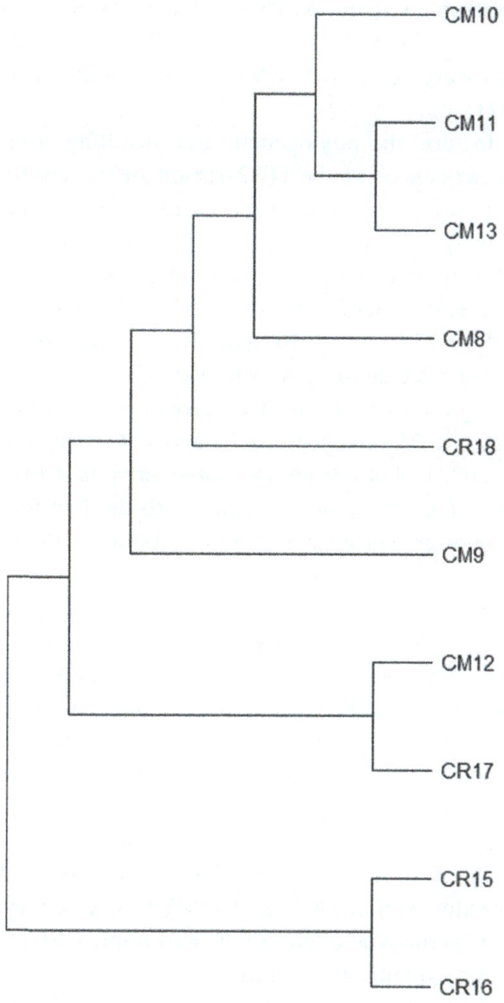

Fig. 8.10 Phylogenetic tree obtained using the Maximum Parsimony (MP) method of the ITS 2 region of cardamom samples from Colombia and Costa Rica, MEGA 11, prepared for this research in 2023

8.6 Maximum Parsimony Analysis of Taxa

The evolutionary history was inferred using the Maximum Parsimony method. Tree #1 out of 8 most parsimonious trees (length = 440) is shown. The consistency index is 0.943182 (0.712644), the retention index is 0.456522 (0.456522) and the composite index is 0.430583 (0.325337) for all sites and parsimony-informative sites (in parentheses). The MP tree was obtained using the Subtree-Pruning-Regrafting (SPR) algorithm (Nei and Kumar 2000) with search level 1, in which the initial trees were obtained by the random addition of sequences (10 replicates). This analysis involved 10 nucleotide sequences. There were a total of 1018 positions in the final dataset. Evolutionary analyses were conducted in MEGA11 (Tamura et al. 2021).

In turn, the phylogenetic tree resulting from the sequences of the ITS2 region of the cardamom samples was analyzed based on origin and phenotype. It should be noted that the sequencing conducted in 2022 and 2023 yielded different results in terms of sequence length, such that CM 12, CR 15, CR 16 and CR 17 have shorter sequences than the other samples.

Figure 8.10 shows the grouping of samples CR 15 (taken in 2003) and CR 16 (taken in 2023)—both from the same area in Costa Rica. This grouping coincides with the fact that there is greater genetic similarity between them, despite the fact that 20 years passed between sampling, since there is so little plant material, and given the long growing cycle and successive reproduction by asexual propagation. For these reasons, their genotypic identity has been maintained.

It can also be observed that sample CM 12 is grouped with sample CR 17. There are no production records for CR 17 from Costa Rica, but in the past, it had exhibited small, round, pale capsules similar to the COMMON category, such as the type of capsule from sample CM 12, which explains its grouping.

Samples CR 18 and CM 9 are genetically closest to each other, which was not expected at this stage of the research. It is worth reminding that sample CR 18 is from Costa Rica, whereas CM 9 is from the in vitro propagation program of material that was believed to have originated in Colombia, but that did not group with the material from Colombia in the initial study, which explains that CM 9 did not come from those plantations as the original sample.

Sample CM 8 is not grouped with any of the other samples, which is explained by the fact that it is genetically improved material, although it is closer to the phenotypes of long, better colored capsules.

The remaining samples, namely CM 10, CM 11 and CM 13, are grouped together. These are categorized as COMMON and INTERMEDIATE, respectively, and the grouping between them allows us to conclude that CM 13 is a genotype similar to the earlier ones.

8.7 Conclusions

Delayed harvesting and damage caused by thrips generate considerable losses for farmers because they result in a final product of poor quality.

While a process that took several years, the improvement program established in Colombia made it possible to obtain improved materials from existing lines. These materials are still found in some plantations of producers associated with the Colombian developer in the land expansion program in southwestern Antioquia.

The sequences of the ITS2 region of the samples from Colombia and Costa Rica show correlation between them, which after the study conducted in 2003 helps to determine which samples from the Colombian improvement program are related to samples found in Costa Rica 20 years later, and which Costa Rican samples are more closely related to certain INTERMEDIATE or COMMON samples from Colombia.

Sample CM 9 does not group with the Colombian samples from which it originated; however, since the material came from in vitro propagation in Costa Rica, due to its proximity in the phylogenetic tree, it seems to be more closely related to sample CR 18 from Costa Rica.

Sample CR 15 from Costa Rica, which was analyzed in the Colombian cardamom improvement program in 2003 and using the ITS2 region in 2023, is grouped in the phylogenetic tree with the current sample from Costa Rica located at EARTH University and labeled CR 16. This could be due to the low genetic variability of materials that has occurred over the past 20 years in the area, the long crop cycle and successive vegetative reproduction of the plants. The same is true for samples CM 12 from 2003 and CR 17 from 2023, of which CM 12 is considered COMMON material in the genetic improvement program in Colombia and CR 17 has no production data, but it is known that the capsule is small, round and pale, which coincides with that category for capsule type.

In 2023, a few plots dedicated to cardamom production were found in Costa Rica; the genetic representation is scarce and they have common materials represented by samples CR 16 and CR 17.

Accession CR 18 found in the canton of Guatuso, Costa Rica, is of a plant with elongated, green capsules. Sample 1 stands out from the rest for having longer capsules and is more closely related to CM 8 in the phylogenetic tree, which is separated from the earlier samples in the same tree.

The cardamom plants relocated to EARTH University allow for research associated with the species and help to conserve the germplasm, which is an important genetic reserve for the area and for the country, as it has genotypes of the species and allows future research to be conducted on the crop.

8.8 Prospects for Cardamom Farming in Both Countries

In the Americas, cardamom is consumed as an exotic spice, so the most profitable economic option is for export. It is necessary to foster alliances to define marketing channels before encouraging planting and offering this alternative to farmers.

The main problem affecting the crop in 2023 is damage by thrips. The incorporation of improved, pest-resistant varieties would be an important option to avoid the excessive use of agrochemicals for pest control, which affects pollinators.

In both Colombia and Costa Rica, there are cardamom seeds with elongated, green capsules of good quality, represented in Colombia by the plants obtained from the developer's genetic improvement program, and in Costa Rica by the variety obtained in the canton of Guatuso in 2023 and conserved at EARTH University for research and preservation of the germplasm.

Acknowledgement To Dr. Sabu K. K., for the opportunity to share cardamom experiences from another continent, to Dr. Mary Mathew for keeping in contact after so many years. In Costa Rica, to the administrative staff at EARTH University, including the Director, the Dean, the Director of Research, the Academic administrator and his team, as well as the Statistic professor, the academic technician of CIDER, the laboratory, and library staff. To producers from Guatuso, for their cooperation and providing plants for research and conservation. In Colombia, I have appreciation and loyalty to the cardamom-producing company, to its owners and employees. To the field workers, from whom I have learned so much over all these years; to the professors and staff at the National University of Colombia; to the reviewers; to my family and "cardamomeritos", for being constant support and motivation.

Annex 1

1. Cardamom from Guatuso, Costa Rica. Left: sample 1, right: sample 3. Photograph taken by the author for this research.
2. Left: cardamom with grain damaged by *thrips*, right: grain past harvest. Photograph taken by the author for this research.

Annex 2

Morphological characteristics of representative accessions, with photographs taken by Tamayo (2003). CM 8, CM 9, CM 10, CM 11, CM 12 and CM 13 are presented.

Annex 3

Left: cardamom plant in Colombia, right: typical spike from the plantation. Photographs of cardamom samples from Colombia in the department of Antioquia, taken by Tamayo (2003).

Annex 4

Evaluation of the length and width measured in millimeters of cardamom samples from the Guatuso Canton, Costa Rica. Statistical analysis of Tukey's method. Made from samples collected in 2023 (Figs. 8.11, 8.12, 8.13, 8.14; Table 8.3).

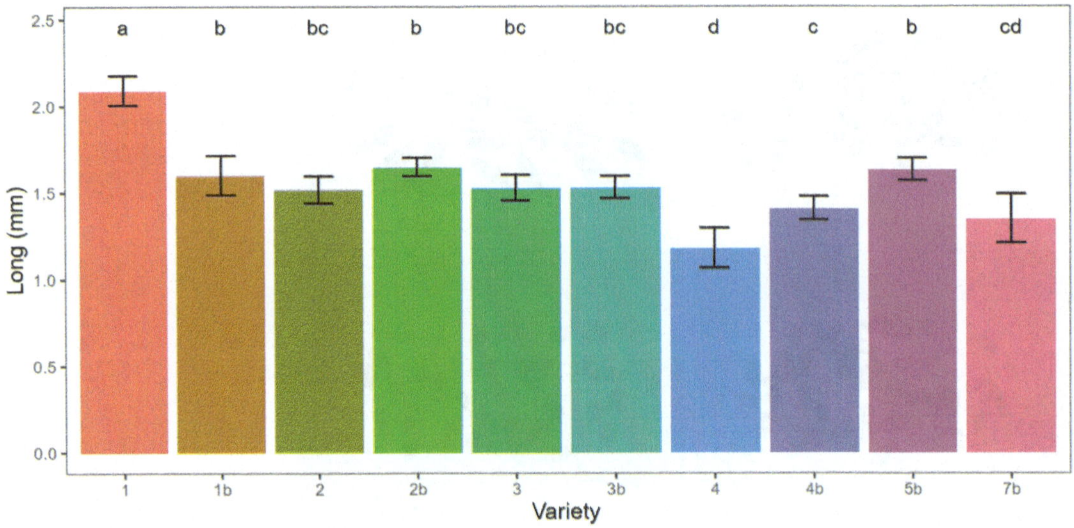

Fig. 8.11 Capsule length data, measured in millimeters in the samples found in Guatuso, Costa Rica, in 2023

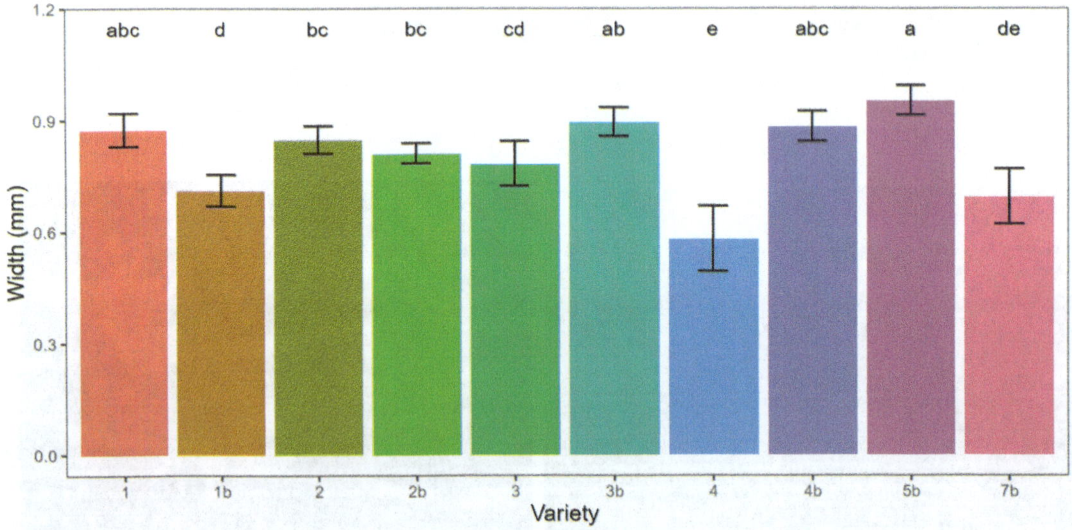

Fig. 8.12 Capsule width data, measured in millimeters in the samples found in Guatuso, Costa Rica, in 2023

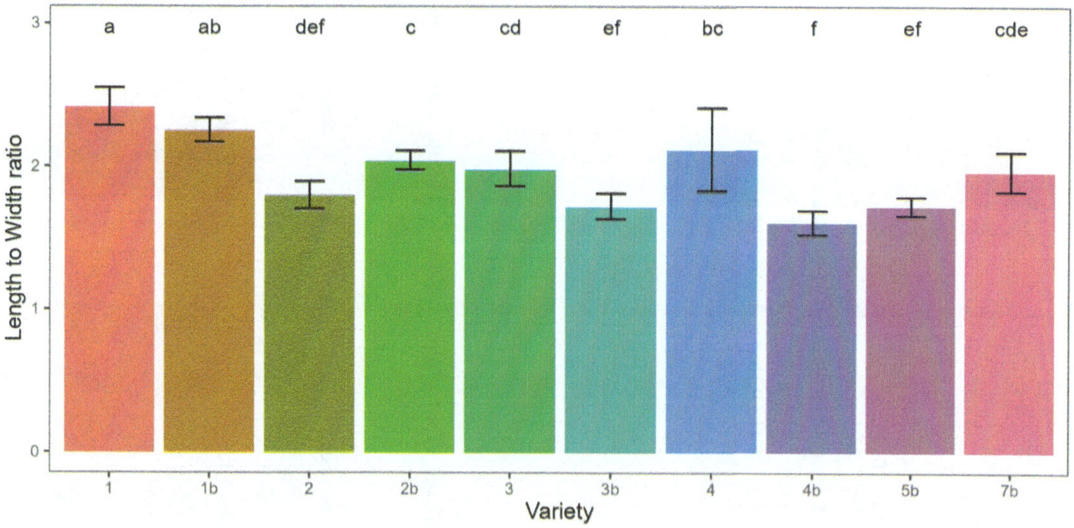

Fig. 8.13 Length/width ratio of capsules in the samples found in Guatuso, Costa Rica, in 2023

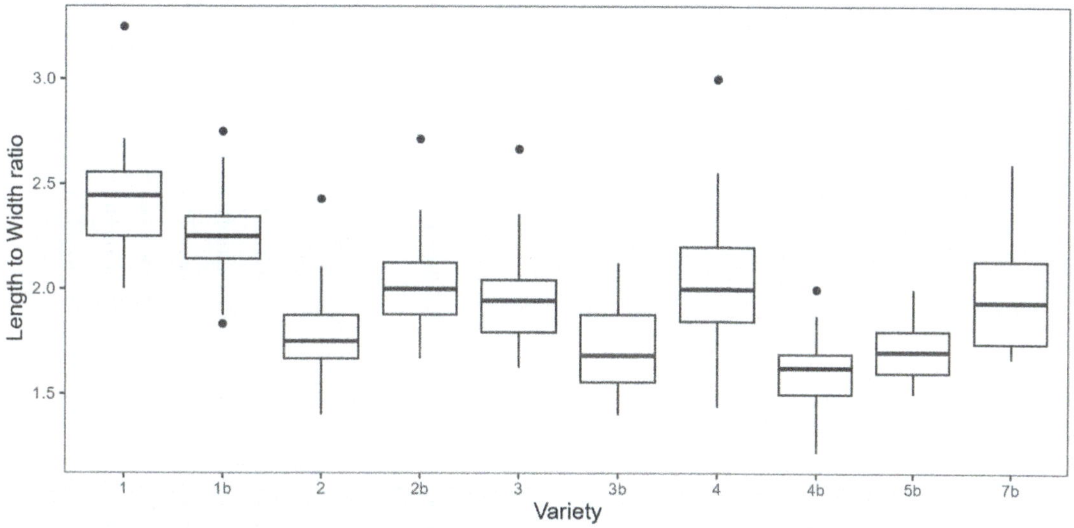

Fig. 8.14 Capsule length/width ratio in the samples found in Guatuso, Costa Rica, in 2023

Table 8.3 Analysis of variables length, width and length/width ratio of capsules in samples from Guatuso, Costa Rica, in 2023

Variety	1	1b	2	2b	3	3b	4	4b	5b	7b
Width	0.87±0.1 abc	0.71±0.11 d	0.85±0.08 bc	0.81±0.08 bc	0.78±0.13 cd	0.9±0.08 ab	0.58±0.15 e	0.88±0.09 abc	0.95±0.09 a	0.69±0.14 de
l_w	2.42±0.29 a	2.25±0.22 ab	1.8±0.21 def	2.04±0.2 c	1.98±0.26 cd	1.72±0.2 ef	2.12±0.48 bc	1.61±0.19 f	1.72±0.15 ef	1.97±0.26 cde
Length	2.09±0.19 a	1.6±0.29 b	1.52±0.17 bc	1.65±0.16 b	1.53±0.16 bc	1.53±0.14 bc	1.18±0.19 d	1.41±0.16 c	1.63±0.14 b	1.35±0.27 cd

Annex 5

Result of the alignment of the sequences of some studied materials. Molecular identification, percentage of coverage and identification, carried out in the CeNAT laboratory, Costa Rica, 2023.

Sample code	Identification BLAST	Access number	% coverage	% identity
CM 8	*Elettaria cardamomum* voucher A.D. Poulsen et al. 2978 internal transcribed spacer 1	MH396864.1	78	97.56
CM 9	*Elettaria cardamomum* voucher A.D. Poulsen et al. 2978 internal transcribed spacer 1	MH396864.1	78	97.40
CM 10	*Elettaria cardamomum* voucher A.D. Poulsen et al. 2978 internal transcribed spacer 1	MH396864.1	79	96.07
CM 11	*Elettaria cardamomum* voucher A.D. Poulsen et al. 2978 internal transcribed spacer 1	MH396864.1	78	98.37
CM 13	*Elettaria cardamomum* voucher A.D. Poulsen et al. 2978 internal transcribed spacer 1	MH396864.1	79	95.95
CR 18	*Elettaria cardamomum* voucher A.D. Poulsen et al. 2978 internal transcribed spacer 1	MH396864.1	80	96.07

Annex 6

Sequences of the ITS2 region of studied materials, carried out in the CeNAT laboratory, 2022.

CM 12
>GACGTCTTGAGAAGTCCATTGAACCTTATCATTTAGAGGAAGGAGAAGTCGTAACAAGGTTTCCGTAGGT
GAACCTGCGGAAGGATCATTGTTGAGAGAGCATTGAATGATGGATGGTTGTGAATGTGTCAACGCGCCCC
TTCCTTGCCCTATGTTGGTGGGCAATTGATCGTAGCTCGGTGCGATCGGCACCAAGGAACAACGAACTTA
GATGCAGAGGGCCCTCGATGTGCGCGGGGAGCTCAATGCGTCGGAGATGCCTCATAATCAAATGACTCTC
GGCAATGGATATCTCGGCTCTTGCATCGATGAAGAACGTAGTGAAATGCGATACTTGGTGTGAATTGCAG
AATCTCGTGAACCATTGAGTCTTTGAACGCAAGTTGTGCCCGAGGCCTTGTGGCCGAGGGCACGCCTGCT
TGGGCGTCATGCATCGTCGCCTTTGCTCCTTGCTTTTGCTGGTGCCAAGCGCGGAAATTGGCCTCTTGTG
CCTCGGGCACAGTCGGTTGAAGAGTGGGTAGTCCGCAGTCGTCGGGCGCGATAGGTGTTGGTCGCCTTG
TGCGTGAACTGATCGTCGTCCCCGTCGTGTTGGGATGAGTCCTCAAGAGACCCTGTGTGAATATGGCGTC
GCATGAAAGCGCCATGTCCGTCATATTGTGGCCCCAAGTCAGGCGAGGCCACCCGCCGAGTTTAAGCATA
TAAATAAGCGGAGGAGGAGAAACTTACAAGGATTCCCCTAGT

CR 15
>ATGCGATACTTGGTGTGAATTGCAGAATCTCGTGAACCATTGAGTCTTTGAACGCAAGTTGTGCCCGAGG
CCTTGTGGCCGAGGGCACGCCTGCTTGGGCGTCATGGCATCGTCGCCTTTGCTCCTTGCTTTTGCTGGTG
CCAAGCGCGGAAATTGGCCTCTTGTGCCCTCGGGCACAGTCGGTTGAAGAGTGGGTAGTCCGCAGTCGTC
GGGCGCGATAGGTGTTGGTCGCCTTGTGCGTGAACTGATCGTCGTCCCCGTCGTGTTGGGATGAGTCCTC
AAGAGACCCTGTGTGAATATGGCGTCGCATGAAAGCGCCATGTCCGTCATATTGTGGCCCCAAGTCAGGC
GAGGCCCACCCGCCGAGTTTAAGCATATCAATAAGCGG

CR 16
>TGCGATACTTGGTGTGAATTGCAGAATCTCGTGAACCATTGAGTCTTTGAACGCAAGTTGTGCCCGAGGC
CTTGTGGCCGAGGGCACGCCTGCTTGGGCGTCATGGCATCGTCGCCTTTGCTCCTTGCTTTTGCTGGTGC
CAAGCGCGGAAATTGGCCTCTTGTGCCCTCGGGCACAGTCGGTTGAAGAGTGGGTAGTCCGCAGTCGTCG
GGCGCGATAGGTGTTGGTCGCCTTGTGCGTGAACTGATCGTCGTCCCCGTCGTTGTTGGGATGAGTCCTCA
AGAGACCCTGTGTGAATATGGCGTCGCATGAAAGCGCCATGTCCGTCATATTGTGGCCCCAAGTCAGGCG
AGGCCCACCCGCCGAGTTTAAGCATATCAATAAGCGGAGGAA

CR 17
>TTTCCGTAGGTGAACCTGCGGAAGGATCATTGTTGAGAGAGCATTGAATGATGGATGGCTGTGAACGTGT
CAACGCGCCCCTTCCTTGCCCCATGTTGGTGGGCAATTGATCGTAGCTCGGTGCGATCGGCACCAAGGAA
CAACGAACTTAGACGCAGAGGGCCCTCGATGTGCGCGGGGAGCTCAATGCGTCGGAGATGCCTCATAATC
TAATGACTCTCGGCAATGGATATCTCGGCTCTTGCATCGATGAAGAACGTAGTGAAATGCGATACTTGGT
GTGAATTGCAGAATCTCGTGAACCATTGAGTCTTTGAACGCAAGTTGTGCCCGAGGCCTTGTGGCCGAGG
GCACGCCTGCTTGGGCGTCATGGCATCGTCGCCTTCGCTCCTTGCTTTTGCTGGTGCCAAGCGCGGAAAT
TGGCCTCTTGTGCCCTCGGGCACAGTCGGTTGAAGAGCGGGTAGTCCGCAGTCGTCGGGCGCGATAGGTG
TTGGTCGCCCTGTGCGTGAACTGATCGTCGTCCCCGTCGTGTTGGGATGAGTCCTCAAGAGAGACCCTGT
GTGAATATGGCGTCGCGTGAAAGCGCCATGTCCGTCATATTGTGGCCCCAAGTCAGGCGAGGCCCACCCG
CCGAGTTTAAGCATATAAATAAGCGGAGGAGGAGAAACTTACAAGGATTCCCCTAGTAACGGCGAGTGAA
CCGGGAT

References

Bhat KV, Lakhanpaul S, Chandel KPS, Jarret RL (1995) Molecular markers for characterization and identification of genetic resources of perennial crops. In: Molecular genetic techniques for plant genetic resources. IPGRI

Dharmadasa M, Nagalingam T, Seneviratne PHM (2008) Sri Lanka. Identification and screening of new generation insecticides against cardamom thrips (*Sciothrips cardamomi*) in cardamom cultivations in Sri Lanka. Research Station, Department of Export Agriculture. Cey J Sci (Bio Sci) 37(2):137–142. http://dl.nsf.gov.lk/bitstream/handle/1/7694/CJS(BSc)-37(2)-137.pdf?sequence=2

Harris SA (1996) Molecular analysis of forest tree biodiversity: a selection of practical protocols. Oxford, England, 47 pp. Modificado por W. Phillips. Enero 23 de 1998

Jacob TK, Senthil Kumar CM, D'Silva S, Devasahayam S, Ranganath HR, Sujeesh ES, Biju CN, Praveena R, Ankegowda SJ (2015) Evaluation of insecticides and natural products for their efficacy against cardamom thrips (*Sciothrips cardamomi* Ramk.) (Thysanoptera: Thripidae) in the field. J Spices Aromat Crops 24(2):133–136

Lang O (1982) La flor, polinización y polinizadores del cardamomo (*Elettaria cardamomum* M.) en Cobán, Alta Verapaz (var. Mysor, Malabar). Tikalia (Guatemala)

Lüttmann NT (1985) El cardamomo y su cultivo. Artemis y Edinter. Talleres de litografías modernas. Guatemala, 83 pp

Nei M, Kumar S (2000) Molecular evolution and phylogenetics. Oxford University Press, New York

Phillips-Mora W (1995) Marcadores De ADN: Teoria, Aplicaciones y Protocolos de Trabajo. Bib. Orton IICA/CATIE

Tamayo AC (2003) Caracterización molecular de materiales seleccionados de cardamom—*Elettaria cardamomum* L. (Matón)—utilizando marcadores AFLP. Universidad Nacional de Colombia, Facultad de Ciencias Seccional Medellín, Trabajo de grado de Maestría en Biotecnología. Medellín/Efe Gómez. https://repositorio.unal.edu.co/handle/unal/84604

Tamura K, Stecher G, Kumar S (2021) MEGA 11: molecular evolutionary genetics analysis version 11. Mol Biol Evol. https://doi.org/10.1093/molbev/msab120

Tapia Delgado KF (2017) Reseña Histórica del Cardamomo—*Elettaria cardamomum* L. (Matón)—en Costa Rica. Tesis Lic Ing Agr. Universidad EARTH, Costa Rica

Thuruthumkara KJ, Senthil Kumar CM, Devasahayam S, D'Silva S, Senthil Kumar R, Biju CN, Praveena R, Ankegowda SKJ (2020) Plant morphological traits associated with field resistance to cardamom thrips (*Sciothrips cardamomi*) in cardamom (*Elettaria cardamomum*). Ann Appl Biol. Indian Council of Agricultural Research (ICAR)-Outreach Programme on Sucking Pests of Horticultural Crops. https://doi.org/10.1111/aab.12592

Universidad EARTH (2023) Estación meteorológica. Datos Ingeniería Agrícola/Datos climáticos

Vos P, Hogers R, Bleeker M, Reijans M, Lee TVD, Hornes M, Friters A, Pot J, Paleman J, Kuiper M, Zabeau M (1995) AFLP: a new technique for DNA fingerprinting. Nucleic Acids Res 23(21):4407–4414

Ana Cristina Tamayo Domínguez graduated in agricultural engineering from the Faculty of Agronomy of the National University of Colombia, and began her work with the cultivation of cardamom in 1993 with a private company in the southwest of Antioquia, in Colombia, in which she participated in the selection process and genetic breeding of the crop, along with the employees at the same company. She completed a Master's degree in Biotechnology, using AFLP molecular markers to evaluate Colombian cardamom samples. Currently, she works at EARTH University of Costa Rica in the area of Applied Genetics and completed her doctorate in education; she has monitored the situation of cardamom cultivation in Colombia and Costa Rica since the 90s and has plants on the EARTH University campus for research and conservation purposes of the species. She has directed numerous degree projects at EARTH University. In 2023, the isolation and sequencing of DNA from the ITS2 region were carried out in samples from Colombia and Costa Rica presented in this chapter.

Functional Regulation of miRNAs in Cardamom

9

F. Nadiya and R. Rajalakshmi

Abstract

MicroRNAs (miRNAs) are long noncoding RNAs of about 18–24 nucleotide length. It is reported in plants, animals and in few viruses. The major function of miRNAs is the negative regulation of targeted mRNAs/genes through the complementary binding in the seed region and eventually leads to target cleavage or translational repression. The differential expression of miRNAs in various conditions confirms the strict functional regulation in various plant developmental processes and stress responses. Moreover, in every plant species, there exists many less conserved miRNAs which shows the functional regulation of miRNAs in almost every aspect of plant biological processes. This chapter explores the regulation of microRNAs in growth, development, yield and drought stress response in small cardamom (*Elettaria cardamomum* Maton). Cardamom is a monocot and 'queen of spices', which is used as a food flavouring agent and also in ayurvedic preparations for digestive problems, mouth ulcers and even for mild depression conditions. There are several miRNAs identified in small cardamom population which exhibited diversity both in its number and position of mature miRNA length and sequences of precursor microRNAs. The differential expression pattern between wild and cultivar pattern showed that 19 miRNAs were differentially expressed and most of the miRNAs were over expressed in cultivar cardamom population. In wild cardamom population, the miR169 was down regulated and miR529 was upregulated suggests that the wild small cardamom populations hold stronger drought stress resistance and increased floral development than its cultivar population. There are 17 miRNAs got differentially expressed in wild cardamom population under drought stress which play significant role in several biological processes to improve drought stress tolerance in plants.

Keywords

MicroRNA · Development · Biotic stress · Up regulation · Next generation sequencing

F. Nadiya (✉) · R. Rajalakshmi
Inter-University Centre for Genomics and Gene Technology, University of Kerala, Karyavattom, Kerala, India
e-mail: fnadiya2009@gmail.com

R. Rajalakshmi
Department of Botany, University of Kerala, Karyavattom, Kerala, India

K. K. Sabu and C. Kole (eds.), *The Cardamom Genome*, Compendium of Plant Genomes,
https://doi.org/10.1007/978-3-031-71829-8_9

9.1 Introduction

MicroRNAs (miRNAs) represent a group of noncoding RNAs of approximately 18–24 nt acts as the critical regulators of plant development processes. Several research states that miRNAs are associated with almost all aspects of plant biology and the regulation is mainly through cleavage and/or translation inhibition of the target mRNAs during or after transcription. miRNA regulations in plants were identified for about 20 years ago as functional regulators in biological processes such as development, maintaining genome integrity, metabolism and environmental stress responses (Pegler et al. 2019a, b). Adopting a combined method, both experimental and in silico approaches is considered as most efficient to identify the functional role of miRNAs in plants. In the initial stages, direct cloning approach was followed to identify miRNAs in plants (Llave et al. 2002; Reinhart et al. 2002). Later, the identification of conserved regions and miRNA stem-loop hairpin structure elucidation through bioinformatics tools lead to the prediction of known as well as conserved miRNAs in new plant species utilizing the already predicted plant miRNA data deposited in the miRNA databases (Jones-Rhoades et al. 2006). From recent years, Next generation Sequencing (NGS) or high throughput-based miRNA profiling is the most accepted and reliable method to functionally predict miRNAs (Lu et al. 2006). NGS-based approaches have several advantages over other methods, it predicts both known, conserved and novel miRNAs in the plant, differential expression patterns in miRNA regulations under specific conditions (Creighton et al. 2009; Fahlgren et al. 2007; Sunkar et al. 2005). Due to this rapid and reliable progress in methods for predicting/identifying plant miRNAs, there was a tremendous increase in the number of miRNAs in the databases from different plant species (Kozomara and Griffiths-Jones 2010; Pegler et al. 2019a, b; Xu et al. 2016). Reports and interpretations from the functional aspects of miRNAs in plants showed that the expression of genes controlled by miRNA is a crucial procedure for plants to get adapted to the biotic/abiotic stress (Sun et al. 2019). Plants when suffer from an abiotic stress strengthens its tolerance to adapt the situation by regulating the expression of specific miRNAs which may be conserved or new miRNAs to adapt to the situation (Jian et al. 2010). The specific miRNA expression may be varied depending on the stress or condition which may either upregulate or downregulate. From the references, it is been proved that miRNAs participate in critical regulation in various fundamental growth and development processes in plants through target key regulatory genes and also in abiotic/biotic stress responses. Similarly, few miRNAs can also be altered to regulate significant agronomic traits in crops to increase its yield and quality (Zhang et al. 2019). Hence, the studies focussing on the stress responsive miRNAs could initiate the development of stress tolerant breeding strategies and also to understand the miRNA-based adaptive mechanism in plants.

Small cardamom (*Elettaria cardamomum* Maton), referred as the 'Queen' of spices is a perennial herbaceous rhizomatous monocot classified under the *Zingiberaceae* family. From time immemorial, this spice is used to increase flavour in food and also as an ayurvedic medicines for mouth ulcers, digestive issues and along with other herbs even for depression. The cultivation of cardamom has been declined for the past few years due to the occurrence of pests and diseases which resulted in a great reduction in total productivity and export. Henceforth, introducing and practising novel breeding methods could produce superior cardamom cultivars with increased pathogen resistance and pest tolerance. The development of such cultivars could compensate the current situation and help farmers to get more profit without investing more into harmful pesticides and insecticide. This chapter discuss the miRNA profiling of cardamom using high throughput sequencing approach and functional regulation of conserved and novel miRNAs associated with agronomical traits and under stress response.

9.2 miRNA Biogenesis

miRNA genes were transcribed into long RNA primary transcripts known as Pri-miRNAs (single stranded) is the first step of miRNA biogenesis. The single stranded Pri-miRNAs are further pleated to form a stem-loop structure (Mallory et al. 2008) and altered into the pre-miRNA (miRNA precursor) and finally into miRNA/miRNA* duplex, one strand is the passenger strand and the other remains as the guide strand. Further, the passenger strand will get degraded by small RNA degrading nucleases and the guide strand will lead the RNA-induced silencing complex (RISC). The major catalytic component of RISC complex is ARGONAUTE1 (AGO1 proteins), which helps the integration of miRNAs and guides the RISC to regulate the target mRNAs. Further, the RISC complex initiates the repression of targeted mRNAs either through translational repression or cleavage (Brodersen et al. 2008; Sunkar and Zhu 2004). This combinatorial regulation is possible in different ways like a miRNA cleaves multiple mRNAs or a single target mRNA is regulated by several different miRNAs (Aukerman and Sakai 2003).

9.3 Regulation of MicroRNAs in Wild and Cultivar Cardamom

Cardamom exists both as wild and cultivar populations, and wild populations are reported to be occurred in isolated patches in their natural habitats; however, the cultivated cardamom is the source of today's commercial product. In Kerala, the wild cardamom populations are sited near tribal areas at the *Idamalayar* forest in the *Ernakulam* district (Kuriakose et al. 2009). A study was done to characterize the differentially expressed miRNAs between wild and cultivar cardamom using Ion proton-based NGS approach (Nadiya et al. 2019). The study was conducted by isolating miRNAs from five high-yielding cardamom cultivars. Two among them are released from Indian Cardamom Research Institute, *Myladumpara*, Kerala (ICRI-5) and from Cardamom Research Centre (Kerala Agricultural University, *Pampadumpara*) (PV2). The remaining three samples were cardamom landraces obtained from the farmers' plots and the control sample is the wild type cardamom collected from the forest area of *Therakkudi*. Usually, the isolation of high integrity RNA from the plant tissues is tedious and hence should be done carefully to get reliable miRNA data. For cardamom, an efficient method was developed to isolate both mRNAs and miRNAs utilizing a commercially available kit (miRNeasy Mini Kit, Qiagen) combined with the modified CTAB method (Nadiya et al. 2015).

High throughput sequencing along with bioinformatics tools and molecular techniques has now become the most routine approach for comprehensive genomics study (Jones-Rhoades and Bartel 2004; Rajewsky and Socci 2004) and the prediction of novel miRNA (Joshi et al. 2010). This approach has been successfully utilized to determine conserved and known miRNAs, novel miRNAs in cultivated and wild populations and the varied expression pattern of miRNAs among the populations and their single targets (Fig. 9.1). Several miRNAs were observed as strong regulators of mRNAs coding significant genes in both cultivated and wild cardamom populations. In both the populations, a total of 42 miRNA families (13 highly conserved miRNA families) were identified. Following a strict miRNA/miRNA* identification criterion, 14 miRNAs were predicted as novel miRNAs which could be considered as small cardamom-specific micro RNAs. However, all the predicted novel miRNA expression level was similar in both the population with very low read count and the conserved miRNAs showed a good read count and were found in both the populations of wild and cultivar cardamom. The differential expression pattern of conserved miRNAs was performed and the results showed that 19 miRNAs (twofold variation) were differentially expressed between the two populations (Table 9.1). Furthermore, the differential expression analysis showed that, most of the

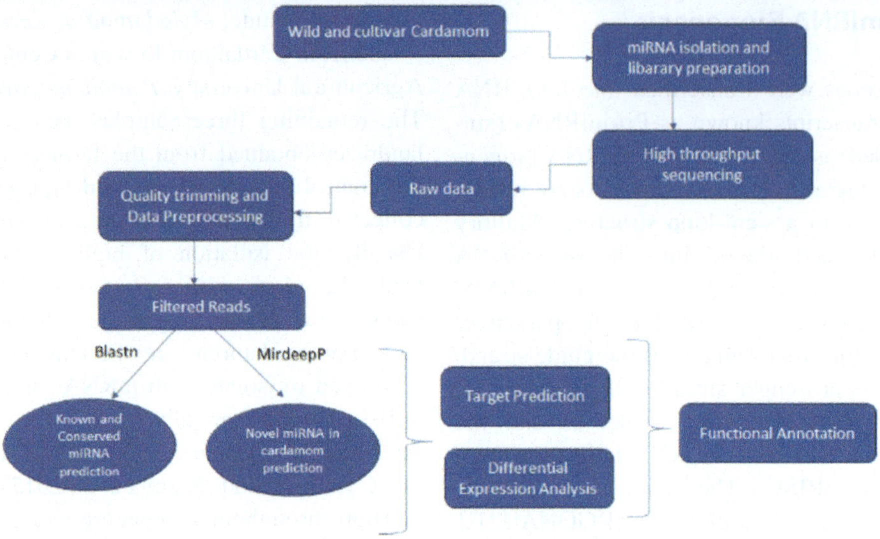

Fig. 9.1 Pipeline used for predicting the regulation of miRNAs in wild and cultivar cardamom

Table 9.1 Differentially expressed miRNAs in wild and cultivar cardamom

miRNA family/miRNA		Fold	Regulation
miR156/159	miR159a	1.8	Upregulation
	miR159d	1.8	Upregulation
miR166	miR166e	1.8	Upregulation
	miR166l	1.7	Upregulation
miR167	miR167h5p	1.8	Upregulation
miR169	miR169d	2.4	Upregulation
	miR169z	1.8	Upregulation
miR171	miR171g	1.8	Upregulation
	miR171b	− 1.4	Downregulation
	miR171i	1.8	Upregulation
miR172	miR172f	1.8	Upregulation
miR319	miR319	1.5	Upregulation
miR393	miR393	1.4	Upregulation
miR398	miR398a	1.8	Upregulation
	miR398b	2.4	Upregulation
miR408	miR408	1.7	Upregulation
miR477	miR477e	2.8	Upregulation
miR5523	miR5523	1.8	Upregulation
miR529	miR529	− 1.4	Downregulation

miRNAs shown upregulation in cultivar cardamom populations. These findings suggest that the upregulation of miRNAs increases the negative regulation of corresponding mRNAs and would influence the stress tolerance level in the cultivar populations. On the other hand, most of the miRNAs identified in cardamom were down regulated in the wild population and there will be an insignificant negative regulation. All these combinatorial regulations affect cardamom cultivars mostly when compared to wild populations making them more susceptible to biotic as well as abiotic stress conditions.

Detailed analysis of miRNAs identified 1168 unique mRNA targets (40 miRNA families) in wild cardamom and 1025 mRNA targets (42 known miRNA families) in cultivar cardamom.

All the predicted miRNAs target several potential genes through both mRNA degradation and translational repression. Few significant findings from the study:

- The miRNAs, miR156/157 and miR529 regulates the genes associated with transcription factor genes like SBP-box (Squamosa promoter Binding Protein-like genes).
- Another miRNA, miR159 targets MYB family transcription factor.
- miR165/166 and miR169 regulate NFYA family genes (Nuclear Transcription Factor Y Subunit A).
- miR171 inhibits and regulates the expression of SCARE-CROW like transcription factor genes.

Few other miRNAs identified in the study plays significant role in the regulation of mRNAs required for the basic functions such as plant defence mechanisms and stress responses.

- mRNAs coding for the enzyme's glutathione-S-transferase, superoxide dismutase is targeted by miR162.
- miR398 regulates genes in hypoxia-responsive family.
- Another highly conserved miRNA 399 regulates genes associated with disease tolerant genes in small cardamom.

There are also several other miRNAs regulating few genes associated with nutrient assimilation and signal transduction (Fig. 9.2).

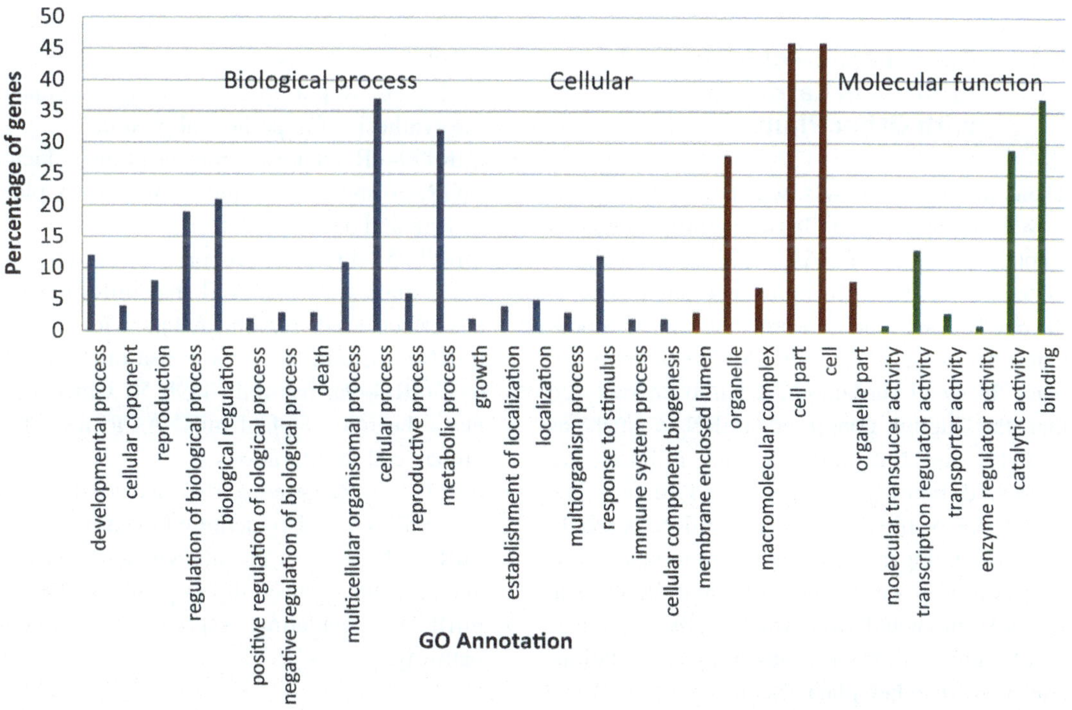

Fig. 9.2 Annotation mRNAs targets identified for conserved miRNAs in cardamom (Nadiya et al. 2019)

The negative gene regulation of miRNAs in cardamom was also studied, and it was observed that only 20% of the regulation is directed through translational repression and the remaining 80% of regulation is through mRNA cleavage. In accordance with previous reports, miRNA profiling in small cardamom also showed increased abundance of few miRNAs such as miR159, miR319 and miR156 in both cardamom populations. Among them, miR159 is more abundant than miR319 and miR156 in both populations. The novel miRNAs identified from cardamom miRNA data showed that they regulate SYT1 (synaptotagmins like proteins) which plays key role in calcium dependent freezing tolerance, mechanical stress response, phytoene desaturase (carotenoid biosynthesis), biosynthesis of secondary cell wall and regulates resistance to bacterial pathogens. Most of these targets were involved in all basic cellular, and metabolic processes, biological regulation, stimulus responses, other developmental and other multiorganismal processes in plants.

9.4 Conservation of Identified miRNAs in Cardamom with Other Plants

There are few miRNAs which has been conserved in monocot or between monocots and dicots. The micro RNAs could be considered as 'conserved between monocots and dicots' is the one bearing a hairpin precursor in the model plant *Arabidopsis* and in any other monocot plant. In small cardamom, it has been reported that the highly conserved miRNAs such as miR319, miR172, miR167 and miR156 are conserved between monocots and dicots. The abundance of highly conserved miRNAs would be representing a significant evolutionary conservation between the monocots and dicots. The miRNAs predicted from small RNA sequences of cardamom showed an evolutionary relationship with in other plant species such as *Oryza sativa* (Szittya et al. 2008), *Triticum aestivum* (Su et al. 2014), *Arabidopsis*, *Zea mays* (Zhang et al. 2009) and *Brachypodium distachyon*. The remaining miRNA families observed in the in the cardamom populations are not conserved between dicots and monocots and are highly specific to monocots. The novel miRNAs predicted could be considered as uniquely expressed in cardamom species and are low abundant in both the populations. Moreover, the miRNAs such as miR528, miR159, miR172, miR156, miR160, miR399, miR166, miR395, miR167, miR390 and miR408 are greatly abundant and highly conserved which regulates the mRNAs required for the basic regulatory mechanisms in all the plants. The miRNA families miR5072, miR5523 and miR535 predicted in cardamom has only one previous report with *Oryza sativa.* Another microRNA annotated as miR5179 form miRbase database has a previous report with both *O. sativa* and *B. distachyon* and not in any other studied monocot plants monocots, in the meantime miR845 was reported in *Brachypodium* and *Arabidopsis*.

Also,

- miR477—Regulates Phenyl-Alanine ammonia Lyase (PAL gene) and increases the susceptibility to diseases.
- miR5279—Reported in Medicago regulates the synthesis of hypothetical protein.
- miR479—Regulates transcription factor StGRAS and plays significant role in plant growth and development.
- miR4995—Previous reports with *Glycine max*, studies also showed that miR4995 has a perfect seed region match with human miRNA hsa-miR-34a-5p, hsa-miR-34c-5p, hsa-miR-449a, hsa-miR-449b-5p which suggests the role of plant miRNAs to modulate human cell proliferations.
- miR3711—Reported earlier as conifer specific miRNA is also identified in cardamom.
- miR6173—Targets the mRNA sesquiterpene synthase and farnesyl diphosphate synthase.
- miR894—Regulating saponin biosynthetic pathway.

The above described miRNA were are not reported in other monocots and they have unique regulatory mechanism and neither conserved among monocots nor between monocots and dicots. Two more miRNAs annotated as miR8175 and miR8579 in the cardamom small RNA data is reported previously only in *A. thaliana*. The study concludes that the miRNAs reported in cardamom are phylogenetically more related to monocots, and there are cardamom-specific miRNAs, miRNAs having seed region similarity with human miRNAs and further more novel miRNAs having unique function. The miRNAs predicted from cardamom populations are validated using 5′RLM-RACE PCR and qRT-PCR. The sequence analysis of RACE PCR products showed that the predicted cardamom miRNAs have specific cleavage sites in their potential target genes.

9.5 Identification and Regulation of Drought Responsive miRNAs in Cardamom

The study on prediction of drought responsive microRNAs in small cardamom using next generation sequencing approach identified a total of 17 miRNAs comprising to 11 miRNA families as drought responsive microRNAs. The study was initiated with the preparation of two small RNA libraries, a water stressed plant and control plant (under normal watering). The plant selected for the study was wild cardamom population from *Idamalayar* forest, *Therakkudy*, *Ernakulam* (Kerala). Among the predicted miRNAs in both control and treated plants, 9 miRNAs have previous reports to be differentially expressed under drought stress.

The miRNA family,

- miR156/159 family members (miR156d and miR159b) were up regulated.
- miR156l was down regulated (Table 9.2).
- Target prediction and annotation of differentially expressed miRNAs showed that that miR156l cleaves the target mRNA Stress Response Suppressor2 (STRS2) which play key role in DEAD-box RNA helicases, binds RNA, involved in drought, salt and cold stress responses (Anjali et al. 2019).

Table 9.2 Drought responsive miRNAs in cardamom

MicroRNA family	MicroRNAs	Fold change/regulation
miR156	miR156d	1.9/up regulated
miR159	miR159b	1.50/up regulated
miR156	miR156l	− 1.25/down regulated
miR166	miR166d-5p	− 2.4/down regulated
miR166	miR166p	− 1.25/down regulated
miR167	miR167f5p	1.68/up regulated
miR169	miR169d	1.64/up regulated

In small cardamom, under drought stress:

- miR169—**upregulated**, which targets Nuclear Factor Y (NFYA) transcription factor subunit A-5 regulating basic developmental process and abiotic stresses response in plants by controlling the stomal opening and closure (Ding et al. 2013).
- The differential expression pattern of remaining 17 miRNAs under drought stress advocates that these miRNAs have significant role in basic biological processes to improve stress tolerance in plants.
- Target annotation showed that several genes involved in water stress response are regulated by miRNAs identified in small cardamom and they are significantly differentially expressed under drought stress.

9.6 Conclusion

The genetic modification of crops through RNAi technology is most powerful to produce better crop yield, which may yield high, can tolerate abiotic and biotic stress factors and can the improve the quality of crops. A number of commercially important crops such as have been improved successfully through miRNA-based technology. We can adopt different methods to develop drought tolerant crops by utilizing miRNA-based genetic engineering technology (Fig. 9.3).

If the miRNA expression is positively correlated with drought stress tolerance traits, we may either over express the corresponding miRNAs or we can knock down the expression of target gene through RNA interference

Fig. 9.3 Different methods to develop transgenic crops through miRNA-based genetic manipulation technology

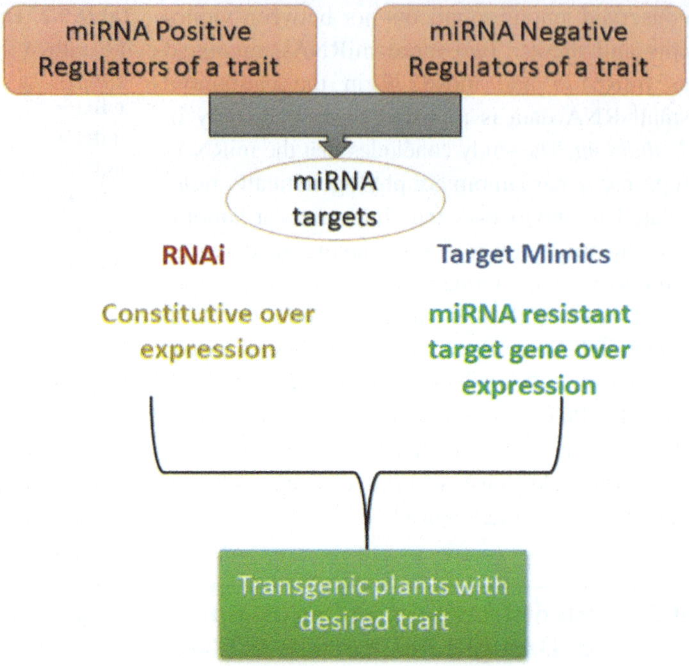

technology. On the other hand, if the expression of miRNA is negatively regulated with the drought tolerant trait, the corresponding target genes need to be over expressed. Another method is to introduce artificial target mimics (a-miRNAs) have to be introduced under specific conditions (Zhou and Luo 2014).

References

Anjali N, Nadiya F, Thomas J, Sabu KK (2019) Identification and characterization of drought responsive microRNAs and their target genes in cardamom (*Elettaria cardamomum* Maton). Plant Growth Regul 87:201–216

Aukerman MJ, Sakai H (2003) Regulation of flowering time and floral organ identity by a MicroRNA and its APETALA2-like target genes. Plant Cell 15(11):2730–2741. https://doi.org/10.1105/tpc.016238

Brodersen P, Sakvarelidze-Achard L, Bruun-Rasmussen M, Dunoyer P, Yamamoto YY, Sieburth L, Voinnet O (2008) Widespread translational inhibition by plant miRNAs and siRNAs. Science 320(5880):1185–1190

Creighton CJ, Reid JG, Gunaratne PH (2009) Expression profiling of microRNAs by deep sequencing. Brief Bioinform 10(5):490–497

Ding Y, Tao Y, Zhu C (2013) Emerging roles of microRNAs in the mediation of drought stress response in plants. J Exp Bot 64(11):3077–3086

Fahlgren N, Howell MD, Kasschau KD, Chapman EJ, Sullivan CM, Cumbie JS, Givan SA, Law TF, Grant SR, Dangl JL (2007) High-throughput sequencing of *Arabidopsis* microRNAs: evidence for frequent birth and death of MIRNA genes. PLoS ONE 2(2):e219

Jian X, Zhang L, Li G, Zhang L, Wang X, Cao X, Fang X, Chen F (2010) Identification of novel stress-regulated microRNAs from *Oryza sativa* L. Genomics 95(1):47–55

Jones-Rhoades MW, Bartel DP (2004) Computational identification of plant microRNAs and their targets, including a stress-induced miRNA. Mol Cell 14(6):787–799

Jones-Rhoades MW, Bartel DP, Bartel B (2006) MicroRNAs and their regulatory roles in plants. Annu Rev Plant Biol 57:19–53

Joshi T, Yan Z, Libault M, Jeong D-H, Park S, Green PJ, Sherrier DJ, Farmer A, May G, Meyers BC (2010) Prediction of novel miRNAs and associated target genes in *Glycine max*. BMC Bioinform 11:1–9

Kozomara A, Griffiths-Jones S (2010) miRBase: integrating microRNA annotation and deep-sequencing data. Nucleic Acids Res 39(suppl_1):D152–D157

Kuriakose G, Sinu PA, Shivanna KR (2009) Domestication of cardamom (*Elettaria cardamomum*) in Western Ghats, India: divergence in productive traits and a shift in major pollinators. Ann Bot 103(5):727–733

Llave C, Kasschau KD, Rector MA, Carrington JC (2002) Endogenous and silencing-associated small RNAs in plants. Plant Cell 14(7):1605–1619

Lu C, Kulkarni K, Souret FF, MuthuValliappan R, Tej SS, Poethig RS, Henderson IR, Jacobsen SE, Wang W, Green PJ (2006) MicroRNAs and other small RNAs enriched in the *Arabidopsis* RNA-dependent RNA polymerase-2 mutant. Genome Res 16(10):1276–1288

Mallory AC, Elmayan T, Vaucheret H (2008) MicroRNA maturation and action—the expanding roles of ARGONAUTEs. Curr Opin Plant Biol 11(5):560–566

Nadiya F, Anjali N, Gangaprasad A, Sabu KK (2015) High-quality RNA extraction from small cardamom tissues rich in polysaccharides and polyphenols. Anal Biochem 485:25–27

Nadiya F, Anjali N, Thomas J, Gangaprasad A, Sabu KK (2019) Deep sequencing identified potential miRNAs involved in defence response, stress and plant growth characteristics of wild genotypes of cardamom. Plant Biol 21(1):3–14

Pegler JL, Grof CPL, Eamens AL (2019a) The plant microRNA pathway: the production and action stages. In: Plant MicroRNAs: methods and protocols, pp 15–39

Pegler JL, Oultram JMJ, Grof CPL, Eamens AL (2019b) Profiling the abiotic stress responsive microRNA landscape of *Arabidopsis thaliana*. Plants 8(3):58

Rajewsky N, Socci ND (2004) Computational identification of microRNA targets. Genome Biol 5:1–35

Reinhart BJ, Weinstein EG, Rhoades MW, Bartel B, Bartel DP (2002) MicroRNAs in plants. Genes Dev 16(13):1616–1626

Su C, Yang X, Gao S, Tang Y, Zhao C, Li L (2014) Identification and characterization of a subset of microRNAs in wheat (*Triticum aestivum* L.). Genomics 103(4):298–307

Sun Z, Su C, Yun J, Jiang Q, Wang L, Wang Y, Cao D, Zhao F, Zhao Q, Zhang M (2019) Genetic improvement of the shoot architecture and yield in soya bean plants via the manipulation of GmmiR156b. Plant Biotechnol J 17(1):50–62

Sunkar R, Zhu J-K (2004) Novel and stress-regulated microRNAs and other small RNAs from *Arabidopsis*. Plant Cell 16(8):2001–2019

Sunkar R, Girke T, Jain PK, Zhu J-K (2005) Cloning and characterization of microRNAs from rice. Plant Cell 17(5):1397–1411

Szittya G, Moxon S, Santos DM, Jing R, Fevereiro MPS, Moulton V, Dalmay T (2008) High-throughput sequencing of *Medicago truncatula* short RNAs identifies eight new miRNA families. BMC Genom 9:593. https://doi.org/10.1186/1471-2164-9-593

Xu M, Hu T, Zhao J, Park M-Y, Earley KW, Wu G, Yang L, Poethig RS (2016) Developmental functions of miR156-regulated SQUAMOSA PROMOTER BINDING PROTEIN-LIKE (SPL) genes in *Arabidopsis thaliana*. PLoS Genet 12(8):e1006263

Zhang L, Chia J-M, Kumari S, Stein JC, Liu Z, Narechania A, Maher CA, Guill K, McMullen MD, Ware D (2009) A genome-wide characterization of microRNA genes in maize. PLoS Genet 5(11):e1000716

Zhang Z, Teotia S, Tang J, Tang G (2019) Perspectives on microRNAs and phased small interfering RNAs in maize (*Zea mays* L.): functions and big impact on agronomic traits enhancement. Plants 8(6):170

Zhou M, Luo H (2014) Role of microRNA319 in creeping bentgrass salinity and drought stress response. Plant Signal Behav 9(4):1375–1391

Whole Genome Sequence and SSR Markers of Small Cardamom (*Elettaria cardamomum*)

10

Sheel Yadav, Ratna Kumari, and Ambika Baldev Gaikwad

Abstract

Small cardamom (*Elettaria cardamomum*), commonly referred to as the "Queen of Spices", is a highly valued spice crop, which is known to have its origin in South India and Sri Lanka. It is one of the oldest spices of the world and is the third most expensive spice after saffron and vanilla. Despite its economic importance, in terms of global market price that this crop commands, very scarce genomic information is available for this crop. This severely limits the feasibility and success of various crop improvement endeavours. To overcome this limitation, the genome of the small cardamom variety, Njallani Green Gold was sequenced. This is the first report for the de novo assembled draft genome sequence of small cardamom. The assembled genome length was 1.06 Gb (gigabases), captured in 8000 scaffolds with a N50 of 0.15 Mb. A total of 68,055 gene models were predicted. The whole genome sequence (WGS) was utilized for the development of 2,27,808 SSR (simple sequence repeat) markers, based on flanking sequence information. A few of these were validated by performing molecular profiling of sixty diverse cardamom accessions. The SSR marker EC 49 was identified to possess the highest polymorphism information content (PIC) of 0.80. Many other highly polymorphic SSR markers, namely EC 1, EC 33, EC 47, EC 89, EC 90, EC 106, EC 183, EC 193, EC 196 and EC 212 were identified with a PIC value>0.60. The SSR markers identified will serve as important genomic resources in cardamom. Many of these can be used for marker-assisted breeding for crop improvement and for germplasm characterization. The information on utilization of the SSR loci for generation of markers has been developed into a public database, the 'cardamomSSRdb'. This database is freely available for use by the cardamom community.

Keywords

Small cardamom · Whole genome sequence · Annotation · SSR marker · Polymorphism information content

S. Yadav · R. Kumari · A. B. Gaikwad (✉)
Division of Genomic Resources, ICAR-National Bureau of Plant Genetic Resources, New Delhi 110012, India
e-mail: ambikabg@gmail.com; ambika.gaikwad@icar.gov.in

K. K. Sabu and C. Kole (eds.), *The Cardamom Genome*, Compendium of Plant Genomes,
https://doi.org/10.1007/978-3-031-71829-8_10

10.1 Introduction

Small cardamom or green cardamom (*Elettaria cardamomum* Maton), often referred to as the "queen of all spices", is immensely valued for its unique flavour and pleasant aroma. It is primarily cultivated in the Southern Indian states of Kerala, Karnataka, and Tamil Nadu. Kerala alone accounts for 87% of the area under small cardamom, highlighting the significant contribution of the state to cardamom production. Historically, small cardamom is believed to have originated in the forests of the Western Ghats of South India and Sri Lankan Central Highlands (Purseglove et al. 1981). Despite its indigenous roots in South India and Sri Lanka, the largest producer and exporter of small cardamom is Guatemala (Preethy et al. 2023). Other countries like Nepal, Indonesia, Costa Rica, Mexico and Tanzania, also contribute significantly to global cardamom production.

Cardamom belongs to the Zingiberaceae family. The family comprises approximately 1600 species (Christenhusz and Byng 2016), which are mostly perennial rhizomatous herbs growing in moist and shady places (Saha et al. 2019). *Alpinia* is the largest genus of the family with more than 300 species (Chaveerach et al. 2008), followed by *Amomum* with 178 species (Thomas and Sabu 2012). Other notable members of this family include the popular spice crops like ginger (*Zingiber officinale*), turmeric (*Curcuma longa*), Javanese ginger (*Curcuma zanthorrhiza*), and Thai ginger (*Alpinia galanga*). Various plant parts of many of these species are widely used in traditional medicine, owing to their therapeutic properties. Many species also serve as sources of natural dyes (Burkill 1966). Rhizomes, in particular are of immense utility as they possess various beneficial medicinal, pharmacological, and nutritional properties (Chan et al. 2011; Akinola et al. 2014). The leaves of a few species like *Kaempferia galanga*, *Curcuma longa*, *Elettariopsis slahmong*, and *Globba marantina* are widely used as flavouring agents. The flowers of many Zingiberaceae species are also visually very attractive and are therefore used for aesthetic purposes.

Small cardamom is a perennial, herbaceous, rhizomatous, monocotyledonous crop. The economic part is the dried, ripe fruit (capsule) which finds great utility for culinary and medicinal purposes. Owing to its pleasant aroma, cardamom is widely used across the different cuisines worldwide, so much so that it serves as a mainstay for the Indian and Mogul cuisines. It is also used as a flavouring agent in European delicacies especially in biscuits and sweet breads. In addition to its culinary use, this spice crop is also known for its medicinal properties, as it is believed to have antioxidant, anti-inflammatory, cholesterol, and blood sugar-lowering properties. A US patent (US 20060147561 A1) has been granted to a poly herbal formulation including cardamom, which can be efficiently used for treatment of diabetes (Nitasha Bhat et al. 2015). Many constituents of cardamom oil have health promoting properties, for instance D-Limonene in cardamom oil, is reported to have tumour suppressing effects and has been found to be effective against colon, mammary, liver, lung, skin, and stomach cancers in rodents (Samir et al. 2015).

Cardamom is one of the oldest spices in the world. Historical records indicate that cardamom cultivation commenced in India around 1803, prior to which cardamom pods were mostly collected from forests (Ravindran and Madhusoodanan 2003). The selection of cardamom for improved agronomic traits began approximately 210 years ago. These practices gradually diminished the presence of wild cardamom and drastically reduced the genetic diversity within the germplasm. The widespread cultivation of specific morpho-ecotypes of the crop led to the development of numerous location-specific varieties. Much like other cultivated plants, wild cardamom species may harbour alleles that can be effectively utilized in crop improvement initiatives. It is estimated that as many as 13 different species of cardamom are reportedly existing in the Western Ghats. It is

Table 10.1 Distinguishing features for the three types of cardamom varieties

Trait	Malabar	Mysore	Vazhukka
Plant structure	At maturity, plants attain a height of 2–3 m	Mature plants attain a height of 3–4 m	Mature plants attain a height of 3–4 m
Panicle type	Prostrate	Erect	Semi-erect
Leaf petiole	Short	Long	Long
Capsule colour at maturity	Pale/golden/yellow	Green	Green
Adaptability and cultivation	600–1200 m, cultivated mainly in Karnataka	900–1200 m, grown in Kerala	900–1200 m, grown in Kerala

Source Prasath and Venugopal (2004)

this genetic diversity which should be effectively utilized for the development of agronomically superior cardamom varieties with greater resilience, in the current climate change scenario. Conserving this diversity, through both in situ and ex situ practices, is therefore of paramount importance. The Indian Cardamom Research Institute (ICRI) at Idukki, Kerala, is the largest repository of cardamom collections with a total of 814 accessions being maintained. These are maintained largely in the field genebanks (ex situ), while in situ conservation is mainly confined to certain protected forest areas such as the Silent Valley Biosphere Reserve in Kerala where a sizable population of cardamom plants are found to be growing in its natural setting. As the conservation of cardamom plants in the field genebanks is difficult due to incidence of diseases like the Katte disease caused by cardamom mosaic virus and rhizome rot due to *Pythium vexans*, the in vitro conservation of cardamom germplasm is considered to be a safer alternative. For this purpose, various protocols have been standardized which involve the use of high-sucrose and slow growth media (Tyagi et al. 2009). Cryopreservation of the seeds at ultra-low temperatures has also been carried out with a reported germination percentage of 80%, for the cryopreserved seeds, after one-year of storage in vapour phase of liquid nitrogen (at 150 °C) (Choudhary and Chandel 1995; Chaudhury and Malik 2004; Ravindran et al. 2004).

The basic chromosome number of cardamom is $x=12$ and $2n=48$ which indicates a tetraploid nature. Being a cross-pollinated crop, there exists huge diversity for various traits like panicle structure, female sterility and cleistogamy (Madhusoodanan et al. 2002; Prasath and Venugopal 2004). Based on fruit size differences, initially the cardamom varieties were classified as *E. cardamomum* var. *major*, which consisted of the wild indigenous types and *E. cardamomum* var. *minor* comprising of the cultivated forms (Wardini and Thomas 1999). The latter includes the cultivars which are classified as Malabar, Mysore, and Vazhukka types which bear prostrate, erect, and semi-erect panicles, respectively. The Vazhukka types are natural hybrids of Mysore and Malabar types. In terms of several qualitative and quantitative traits, the Malabar variety is relatively superior compared to the other varieties. These traits include enhanced drought tolerance (Sudharshan et al. 1991), higher yield (Padmini et al. 2000), early maturity and thus suitable for high-density planting (Madhusoodanan et al. 2002), higher flavonoid content in capsules (Govindarajan et al. 1982), etc. The major differences between these varieties are listed in Table 10.1.

10.2 Genetic Diversity

Genetic diversity is the key to crop improvement. Ramanatha Rao and Hodgkin (2002) describe genetic diversity as the sum total of genetic characteristics within any species or genus. Hajjar et al. (2008) demonstrated that with an increase in genetic diversity, there is an improvement in pest and disease management which further provides an opportunity for improvement of the species. The varieties with little or no genetic diversity are more susceptible or vulnerable to damage due to biotic and abiotic stresses (Salgotra and Chauhan 2023). The presence of diverse accessions within a species allows breeders to develop new and improved cultivars with enhanced resilience to the ever changing environment. The development of such improved cultivars is the need of the hour in order to achieve the goal of food and nutritional security in future. The monoculture farming practices which involve widespread cultivation of a select few varieties over a huge area, along with the rapid environmental deterioration have led to a drastic loss of genetic diversity. This erosion of genetic diversity poses a severe threat to the world's food and nutritional security. The traditional landraces and farmer's varieties are a highly valuable reservoir of superior genes or alleles which can be transferred in the cultivated gene pool of a crop in order to develop agronomically superior cultivars. Therefore, we must conserve this diversity, a goal which is enshrined as one of the seventeen Sustainable Development Goals (SDGs) of the United Nations (UN), which aims at achieving food security and eradicate hunger (https://www.un.org/sustainabledevelopment/sustainable-development-goals/).

As very few genomic resources are available in the Zingiberaceae family, due to lack of DNA sequence information, most genetic diversity studies for this family have utilized morphological traits and random or non-sequence specific DNA-based molecular markers. The various DNA-based molecular markers that have been used for diversity analysis in various Zingiberaceae species are presented as Fig. 10.1.

India, being the centre of origin, has rich diversity in small cardamom. This diversity must be integrated for breeding cardamom varieties with improved agronomic performance. Presently, the cardamom improvement programmes are aimed at development of high yielding, drought tolerant and heat resistant varieties, and varieties resistant to clump rot and various viral diseases (Jose et al. 2022). To achieve these goals, one needs to identify and select genotypes possessing the desirable traits. Wild cardamom genotypes are rich in novel alleles that could positively contribute to the genetic improvement of the cultivars (Nadukeri et al. 2020). However, for a perennial crop like cardamom, cultivar selection can be extremely challenging and difficult, as morphological discrimination can be done only beyond a particular growth stage, when the phenotype is clearly evident. Hence, the use of DNA-based molecular markers for cultivar identification is a reliable alternative which can be used to supplement morphological descriptors. There are many

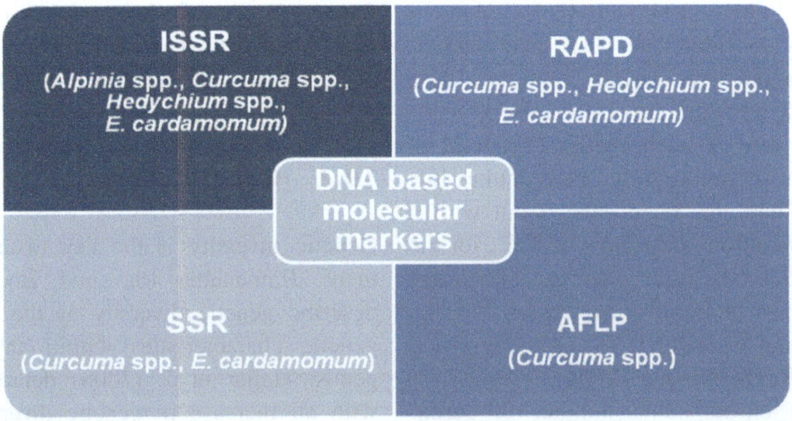

Fig. 10.1 Various DNA-based molecular markers used for diversity analysis for the Zingiberaceae species (data taken from Saha et al. 2019). ISSR: inter simple sequence repeat; RAPD: random amplified polymorphic DNA; SSR: simple sequence repeat; AFLP: amplified fragment length polymorphism

advantages associated with the use of molecular markers for genetic diversity assessment. Unlike morphological descriptors, which are highly influenced by environment, molecular markers have high reproducibility and amenability to automation. Also, these are free from any environmental fluctuations and are therefore neutral in nature. The utilization of molecular markers for genetic diversity analysis paved the way for genetic mapping and marker-assisted selection (MAS) in breeding (Lapitan et al. 2007). Therefore, many studies on genetic diversity have been reported to use both morphological and molecular markers simultaneously.

In small cardamom, In the absence of whole genome sequence (WGS) information of small cardamom until now, very few DNA-based markers were available. Most of the studies which aimed at estimating the genetic diversity in the cardamom germplasm relied on use of morphological traits or random molecular markers like RAPD and ISSR markers. In particular, the ISSR markers have been most widely used for diversity analysis in cardamom, studying genetic relatedness and trait identification (Jose et al. 2013; Anisha et al. 2020; Prasannan and Jose 2021). Variety specific ISSR markers have also been identified in cardamom (Jose et al. 2013), and recently, a Sequence Characterized Amplified Region (SCAR) marker has also been developed from the ISSR sequence information, which allows identification of the Malabar variety (Jose et al. 2022). The absence of sequence-based markers greatly limits the scope of diversity analysis and mapping. The RAPD and ISSR markers as such are relatively difficult to score and are crippled with low reproducibility. The identification of SSRs would broaden the scope of diversity assessment as these are abundantly present in the genome, are highly reproducible and amenable to high-throughput profiling. Table 10.2 details the studies carried out to analyse the genetic diversity and stability in cardamom. Based on the diversity analysis, the germplasm accessions are being grouped into different clusters or groups. This eventually allows identification of genetically diverse accessions. Hybridization between such diverse accessions is believed to result in an increased vigour (hybrid vigour) and fertility for the hybrid developed. This is more relevant for the cross-pollinated and asexually reproducing species, where most of the improved varieties developed are either hybrids or synthetics/composites, which derive their superior performance due to the phenomenon of hybrid vigour.

10.3 Whole Genome Sequence (WGS)

The genomic sequence data, in the form of WGS, represents the ultimate level of genetic information. It allows unravelling of the complexities of all life forms in the world. Sequencing the genome therefore remains a fundamental and integral step in understanding the biological diversity on earth. However, there are certain challenges which make sequencing plant genomes difficult such as presence of vast proportion of repetitive sequences, rampant polyploidy, and heterozygosity. Despite these challenges, genomes of more than 798 land plant species have been sequenced (Marks et al. 2021).

DNA sequencing based on the short read Sanger sequencing technology was expensive. In the early 2000s, the Roche 454 and Illumina platforms ushered in the new technological advances in next-generation sequencing (NGS) technology, which led to a significant increase in sequencing capacity and lowered the sequencing costs substantially. These led to paradigm shifts in genome sequencing technologies. Over the years, the cost involved in genome sequencing has decreased significantly. With the advances in NGS technology and new computational tools, it has now become possible to sequence and assemble the genome of any species at an unprecedented pace and precision. There are many reviews available which elaborate and comprehensively describe the NGS sequencing technologies and how these have revolutionized the field of plant breeding and genomics (Govindaraj et al. 2015; Swarup et al. 2021; Uhlen and Quake 2023).

For the Zingiberaceae family, the WGS information is available for four different species,

Table 10.2 Reports on use of morphological and molecular markers for characterization of small cardamom germplasm

S. No.	References	Plant material	Methodology	Significant findings
1.	Prasath and Venugopal (2004)	310 accessions	16 different agronomic traits utilized	The accessions clustered into three distinct groups
2.	Babu et al. (2012)	96 accessions	RAPD, PCR-RFLP, ISSR	Narrow genetic base for varieties originating from the same centre
3.	Setyawan et al. (2014)	10 accessions	Isozymes	Efficacy of isozymes for estimating species diversity was established
4.	Anjali et al. (2015)	18	EST-SSRs from *C. longa*	The EST-SSRs developed were successfully amplified in cardamom and could differentiate all the accessions used in the study
5.	Hrideek et al. (2015)	10 accessions	63 morphological traits	Highest heritability was observed for the trait leaf length and highest genetic advance was observed for the trait number of tillers/clump
6.	Cyriac et al. (2016)	20 accessions	SSR	Genotype-specific marker profiles were identified, additionally the developed markers showed high transferability (71.4%) across related Zingiberaceae species
7.	Anjali et al. (2016)	30 accessions	ISSR	Highly diverse accession, C53 was identified among the collection
8.	Phadnis and Peter (2017)	11 accessions	RAPD	The genotypes were clustered into two major clusters based on geographical location and plant type
9.	Sakthipriya and Sabu (2018)	18 accessions	EST-SSRs of ginger	EST-SSRs of ginger were successfully amplified for cardamom and the wild type accessions could be differentiated from cultivated accessions
10.	Anisha et al. (2020)	13 accessions	ISSR	Various morphological traits and ISSRs were utilized and the genetic diversity among the accessions was estimated
11.	Nadukeri et al. (2020)	67 accessions	Agronomic traits	The trait number of leaves per tiller was most variable as it had high CV (coefficient of variation) values
12.	Heryanto and Syukur (2021)	100 accessions	Agronomic traits (both qualitative and quantitative)	The accessions were grouped into three distinct clusters and highest variability was observed for the trait number of tillers

(continued)

Table 10.2 (continued)

S. No.	References	Plant material	Methodology	Significant findings
13.	Prasannan and Jose (2021)	10 accessions	ISSR	ISSR markers which distinguished high yielding varieties from low yielding varieties were identified
14.	Jose et al. (2022)	50 accessions	ISSR	A sequence characterized amplified region (SCAR) marker specific to Malabar variety was developed
15.	Preethy et al. (2023)	67 accessions	Agronomic traits recorded over a period of three consecutive years	High diversity among cardamom accessions originating from the evergreen forests of Western Ghats
16.	Gaikwad et al. (2023a, b)	60 accessions	SSR	A high degree of population admixture was observed which indicated the prevalence of cross-pollination in the species

namely *Zingiber officinale* (Li et al. 2021), *Curcuma longa* (Chakraborty et al. 2021), *Curcuma alismatifolia* (Dong et al. 2022), and *Elettaria cardamomum* (Gaikwad et al. 2023a). For *Z. officinale*, the genome was assembled into two haplotypes, designated as haplotype 1 and haplotype 0, the ginger genome being highly heterozygous. For *C. longa*, the draft genome (Chakraborty et al. 2021) has been improved to yield chromosome scale, assembled genome (Yin et al. 2022). The details of the assembly for the four species are given in Table 10.3.

10.3.1 WGS of Cardamom

The WGS of one of the most popular, Indian variety of cardamom, Njallani Green Gold has been assembled using hybrid assembly strategy. This is a high yielding variety of cardamom which is very popular among cardamom farmers of Kerala and has since its development revolutionized the cardamom cultivation in the country.

10.3.1.1 Sequencing and Assembly

Hybrid assembly of long and short read data is frequently adopted. This yields scaffold level assembly while overcoming the relatively low accuracy of the data generated from long read sequencing. The small cardamom genome has been sequenced using both the long (Oxford Nanopore-ONT) and short read sequencing chemistries (Illumina, 10x Chromium) were utilized for sequencing. The Nanopore sequencing generated a total data of 141.90 gigabases (Gb). After data correction, which involved retention of only good quality reads, the filtered reads were assembled de novo and an assembled sequence of 1.06 Gb was generated. More than 75% of the genome was assembled at a sequencing depth was 108.46 X. The assembled genome was subsequently polished using the Illumina paired-end (PE) data. The Illumina sequencing chemistry generated a total of 105.11 Gb data. The 10 X GemCode sequencing data along with the Nanopore de novo assembly was used for scaffolding. The final assembly consisted of 8000 scaffolds with a scaffold N50 of 0.15 Mb. There was a progressive increase in the N50 value which indicated the improvement in assembly as a result of polishing and the use of 10x chromium sequencing data for scaffolding. The length of the assembled genome was close to the predicted genome size of 1.4 Gb (Anjali et al. 2016). The workflow followed for the genome sequencing and assembly is described as Fig. 10.2.

Table 10.3 Members of Zingiberaceae family with the WGS available

Species	Sequencing chemistries	Assembled genome size	Number of annotated genes	References
Z. officinale	PacBio + Illumina + Hi-C	Haplotype 1–1.53 Gb Haplotype 0–1.51 Gb	Haplotype 1–39,217 Haplotype 0–38,090	Li et al. (2021)
C. longa	Oxford Nanopore + 10x Genomics	1.02 Gb	50,401	Chakraborty et al. (2021)
	PacBio + Hi-C	1.11 Gb	49,612	Yin et al. (2022)
C. alismatifolia	PacBio + Hi-C	991.3 Mb	33,902	Dong et al. (2022)
E. cardamomum	Oxford Nanopore + Illumina + 10x Genomics	1.06 Gb	68,055	Gaikwad et al. (2023a, b)

Benchmarking Universal Single-Copy Orthologs (BUSCO) is the tool for assessing the completeness of an assembled genome. The analysis provides a score called the BUSCO score which depicts the presence of single-copy genes, typically present in single-copy across major lineages, in the assembled genome. If an assembled genome has full-length copies of single-copy BUSCO genes, it is expected to be complete, capturing a significant amount of the expected gene content. Therefore, a high BUSCO score is desirable. For the cardamom genome, more than 72% of the genes were found to be complete (C) with more than 99% single-copy genes, on analysis with 1375 total BUSCO groups for *Arabidopsis thaliana* (Gaikwad et al. 2023a).

10.3.1.2 Genome Annotation

Genome annotation enables broader utilization of the genome sequence within the scientific community by furnishing meaningful information about the genome's gene count, repetitive elements, and regulatory and non-coding regions. Genome annotation, defined as the identification of functional elements throughout the genome sequence, imparts significance to the genetic information. The Gene Ontology (GO) and Kyoto Encyclopaedia of Genes and Genomes (KEGG) pathway analyses are widely employed for gene enrichment analysis based on annotation. GO analysis establishes associations between genes and the three GO terms: Biological Processes (BP), Molecular Functions (MF), and Cellular Components

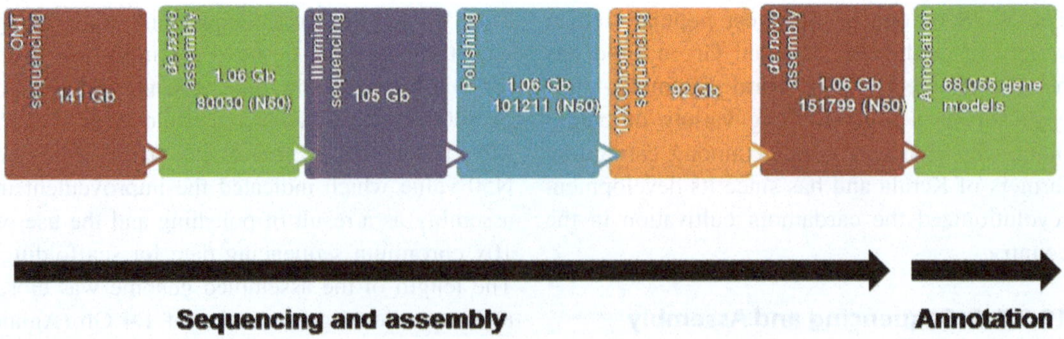

Fig. 10.2 Workflow enlisting the various steps followed for sequencing and assembly of the cardamom genome

(CC), classifying genes within these categories. Meanwhile, KEGG pathway analysis facilitates the identification of gene enrichment in diverse metabolic pathways. This is achieved by linking a set of genes in the genome to a network of interacting molecules in the cell, forming a biological pathway or complex that illustrates a higher-order biological function (Kanehisa and Goto 2000).

A total of 68,055 gene models have been identified in the cardamom genome. Among the non-protein coding genes, there were 478 tRNA genes and 1852 other small RNA genes. Notably, repetitive sequences constituted a substantial portion, making up to 71% of the assembled genome (Gaikwad et al. 2023a). Retrotransposons are reported to have played a significant role, contributing as much as 47% to the repetitive elements. Within the realm of retrotransposons, LTR retrotransposons constituted 46% of the assembled sequence, with approximately 36% being Copia elements and around 9% being Gypsy elements.

The GO analysis unveiled enrichment of the identified genes in three key categories: cellular component, molecular function, and biological process. Within the cellular component GO category, the term "membrane" (GO: 0016020) stood out as the most abundant. Similarly, in the molecular function GO category, the term "binding" (GO: 16390) predominated. In the biological process GO category, the terms "cellular process" (GO: 0009987) and "metabolic process" (GO: 0008152) were prominently represented. Additionally, KEGG enrichment analysis highlighted the enrichment of genes participating in metabolic pathways and the biosynthesis of secondary metabolites. Notably, species within the Zingiberaceae family are recognized for their richness in secondary metabolites, which have been extensively utilized in traditional medicine. The identification of genes involved in the biosynthesis of these compounds holds significance for understanding the underlying genetic mechanisms of secondary metabolite production in cardamom.

10.3.1.3 Comparative Genomic Analysis

To gain insights into the evolution of cardamom, a comparative analysis was conducted with the genomes of other species. The highest similarities were observed between cardamom and the wild banana species *Musa acuminata*, with 7265 orthologous genes. This was followed by the banana species *Musa balbisiana*, which showed 4996 orthologous genes (Fig. 10.3).

In total, 3708 gene families were identified in the cardamom genome (sum of the figures enclosed in red space). The highest number

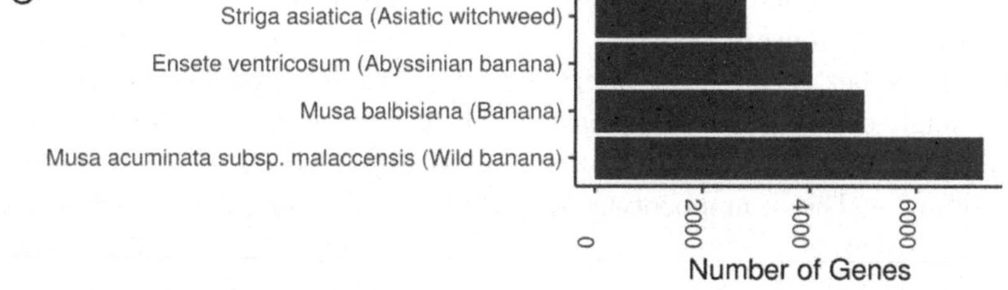

Fig. 10.3 Orthologous genes identified between cardamom and different plant species

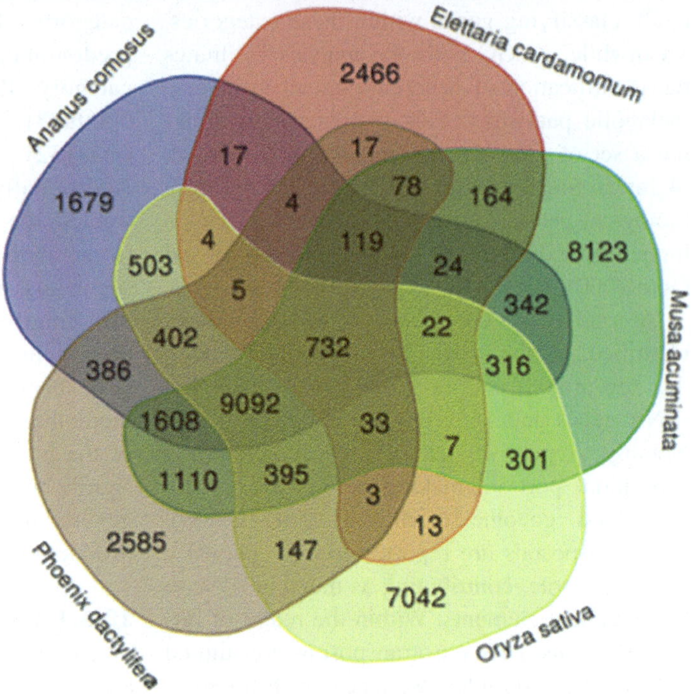

Fig. 10.4 Venn diagram representing the clusters of gene families in cardamom shared with *Ananas comosus*, *Musa acuminata*, *Oryza sativa*, and *Phoenix dactylifera*

of gene families, 31% (1179), were conserved between cardamom and *M. acuminata* (Fig. 10.4), as was expected. This is inferred from the shared number of gene families between the red and green enclosures. This finding aligns with their taxonomic classification, as both *M. acuminata* and cardamom belong to the same order of Zingiberales (D'Hont et al. 2012). Moreover, a total of 732 gene families were conserved among all five species analysed, namely *Ananas comosus* (Family: Bromeliaceae), *E. cardamomum* (Family: Zingiberaceae), *M. acuminata* (Family: Musaceae), *Oryza sativa* (Family: Poaceae), and *Phoenix dactylifera* (Family: Arecaceae), while 2466 gene families were specific to cardamom as is evident by their exclusive presence within the cardamom genome.

The phylogenetic relationship between cardamom and thirteen other species was inferred based on the single-copy protein sequences (Fig. 10.5). These species belong to diverse taxonomic groups. The monocotyledons (Commelinids) included *A. comosus*, *Cocos nucifera*, *E. cardamomum*, *M. acuminata*, *M. balbisiana*, *O. sativa*, *Phoenix dactylifera*, *Sorghum bicolour*, and the eudicots consisted of *Arabidopsis thaliana*, *Glycine max*, *Populus deltoids*, *Solanum lycopersicum*, and *Vitis vinifera*. The basal angiosperm *Amborella trichopoda* was used as an outgroup species. The phylogenetic analysis clearly delineated the Commelinids from the eudicots. Highest genetic similarity was present between the genomes of cardamom and *M. acuminata*. This finding aligns with previous studies that have consistently shown a close phylogenetic relationship between Zingiberaceae species like *C. longa*, *Z. officinale*, and *C. alismatifolia*, and species from the Musaceae family, particularly *M. acuminata* and *M. balbisiana* (D'Hont et al. 2012; Dong et al. 2022). Furthermore, our observations indicated that Arecales species such as *P. dactylifera* and *C. nucifera* are more closely related to Zingiberales (including *Musa*) than to Poales, consistent with previous reports (D'Hont et al. 2012; Yin et al. 2022). This highlights the intricate and fascinating evolutionary relationships among plant families. The analysis of gene family expansion and contraction provides key insights into the evolutionary history of an

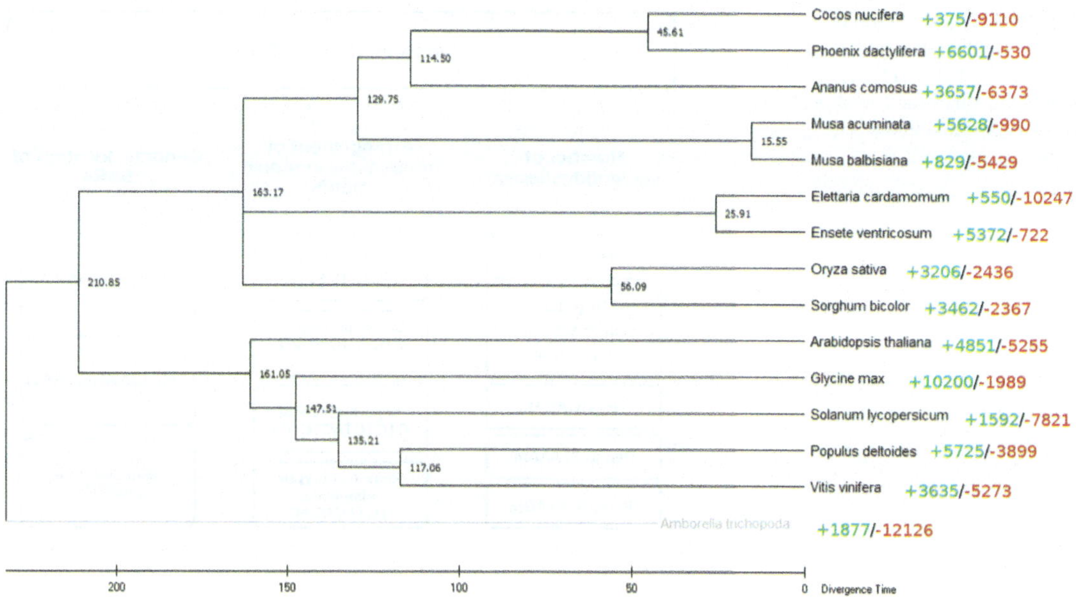

Fig. 10.5 Phylogenetic tree of cardamom with 13 other species based on the single-copy protein sequences. *Amborella trichopoda* was used as an outgroup species. The coloured figures represent CAFÉ-based estimates of gene family expansions (+) and contractions (−). The scale at the bottom depicts the divergence time in million years ago (Mya) with individual figures given at the branches

organism. The relative expansion and contraction of certain gene families allows the organism to adapt to different ecologies. The expansion and contraction of gene families is a result of gene duplication and gene loss events which are both very dynamic phenomena (Rogers et al. 2022). Natural selection also shapes the spectrum of gene family evolution. This analysis revealed that 550 gene families show expansion in cardamom and 10,247 gene families were contracted.

10.4 Simple Sequence Repeats (SSRs)

Simple sequence repeats (SSRs), also known as microsatellites, are short tandem repeated motifs that may vary in the number of repeats at a given locus in the genome (Tautz 1989). Polymorphism identified through these markers is based on PCR reaction, which allows detection of any variations at a particular locus, arising due to differences in the number of repeats at the locus. Microsatellite repeats are classified in various ways. According to their location in the genome, they can be classified as nuclear (nuSSR), mitochondrial (mtSSR), or chloroplast SSRs (cpSSR) (Fig. 10.6). Other ways of SSR classification include classification based on the number of bases in the core repeat, i.e. mono-, di-, tri-, tetra-, penta-, or hexa-SSRs, and into perfect, imperfect, and compound SSRs, which display perfect repetitions, interruption with novel nucleotides, and two or more tandem motifs, respectively (Ahmad et al. 2018). Although SSRs are usually defined as repeats of 1- to 6-bp motifs, heptanucleotide and octanucleotide SSRs have also been identified (Biswas et al. 2014).

Compared to other molecular markers, SSR markers have many advantages. These are their relative abundance, genetic co-dominance, multi-allelic nature, and ease of scoring (Powell et al. 1996). Due to these advantages, SSR markers have been used, over the past few years,

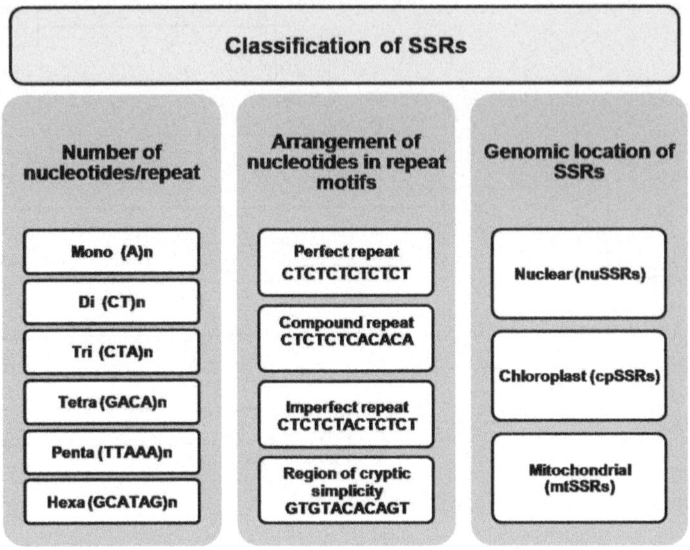

Fig. 10.6 Classification of microsatellites (data taken from Kalia et al. 2011). n = the number of times the core repeat motif is repeated in the SSR region

for a variety of purposes like genetic diversity analysis (Emanuelli et al. 2013; Ren et al. 2014; Kumari et al. 2019), parentage assessment or hybridity testing (Malysheva et al. 2003), species identification (Shirasawa et al. 2013), and gene tagging and genetic mapping (Lu et al. 2013). These are recommended by UPOV (International Union for the Protection of New Varieties of Plants) for plant variety identification and database construction (Zhou et al. 2020). It is due to these advantages that SSR markers have now been identified in as many as 112 different plant species (Song et al. 2021).

Since these are sequence-based markers, some form of DNA sequence information is required for their development. Conventionally, the SSR markers were developed based on microsatellite enriched library construction (Saxena et al. 2015). However, these methods are time consuming and relatively expensive. The development of expressed sequence tag-SSR (EST-SSR) markers is a fast and relatively inexpensive way of SSR marker development, but it is limited to species for which this type of information is available (Gupta et al. 2003; Feng et al. 2016). Databases such as the Genbank (www.ncbi.nlm.nih.gov/genbank/) hold a huge number of EST sequences, gene, and cDNA clones. Similarly, the TIGR Institute (The Institute for Genomic Research, http://www.tigr.org/tdb/e2k1/plant.repeats/) possesses an EST database, which provides EST sequences for many cultivated plant species of economic importance. Alternatively, SSRs of one species can be utilized successfully for molecular profiling of a related species (Awasthi et al. 2017). With the availability of cost-effective, NGS sequencing technologies one can identify SSRs across the genome by sequencing the genome, at a low coverage. This has lately been a method of choice for SSR marker development in many plant species for nuclear (Kumari et al. 2019; Bhattarai et al. 2021; Pei et al. 2023; Sari et al. 2023; Zhao et al. 2023) as well as chloroplast SSRs (Gaikwad et al. 2023a; Kaila et al. 2016).

For the Zingiberaceae family, SSRs have been identified in many species. These have been developed through different approaches. EST-SSRs were developed in *Zingiber officinale* and later used successfully for other Zingiberaceae species like *Zingiber zerumbet*, *Hedychium spicatum*, *Curcuma longa*, *C. amada*, *C. aeruginosa*, *C. aromatic*, and *C. angustifolia* (Awasthi et al. 2017). Taheri et al. (2019) used de novo transcriptome assembly through Illumina sequencing to mine SSRs for *C. alismatifolia*.

For identification of SSRs, several programmes are available. Varshney et al. (2005) have enlisted all the available tools which can

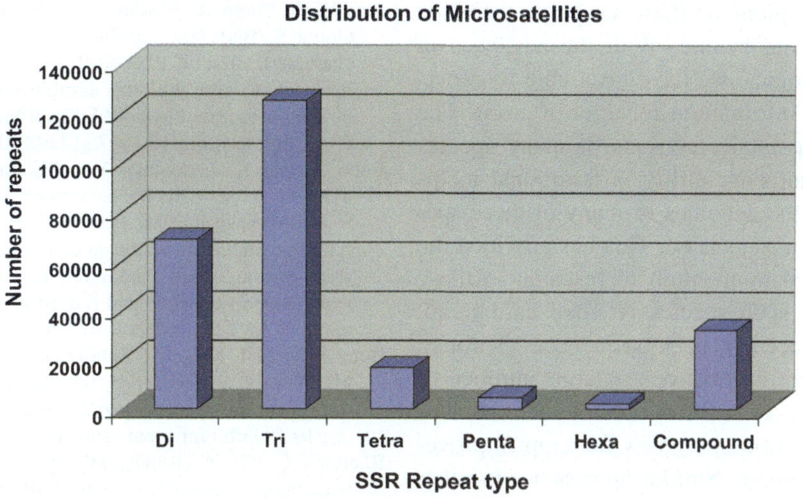

Fig. 10.7 Distribution of different types of SSRs identified in the cardamom genome

be used for SSR identification. Among all these tools, MISA (MIcroSAtellite identification tool, Beier et al. 2017) is a widely used tool, with more than 900 citations. For SSR identification through this tool, the FASTA sequences (entire genome or individual sequences) are used as input files. The core motif length (1–6 nucleotides) and the number of times it is tandemly repeated are usually kept as default parameters.

For small cardamom, utilizing the WGS (Gaikwad et al. 2023a), a total of 250,571 SSRs were identified of which 218,270 (87%) were perfect SSRs and 32,301 (12%) were compound SSRs. For the perfect SSRs, trinucleotides were most abundant (125,329), followed by dinucleotides (68,831), tetranucleotides (16,794), and pentanucleotides (4936). The hexanucleotide repeats were least in number (2380) (Fig. 10.7). All the information regarding the identified SSR markers has been deposited in a database (Green cardamom db). This database can be utilized to search for microsatellite repeats and their corresponding primers for PCR-based amplification at the identified SSR loci in cardamom (http://www.nbpgr.ernet.in:9092/).

Sixty SSR markers were selected for performing further validation studies. For this purpose, these were used for molecular profiling of sixty diverse cardamom accessions. A total of 874 alleles were detected in this analysis. The average number of alleles amplified were 14.57 alleles per locus. The number of alleles ranged from a minimum of 4 to a maximum of 30. The marker EC 49 had the highest value for polymorphism information content (PIC), i.e. 0.80, while the marker EC 12 had the least PIC value of 0.08. A set of SSR primers with high PIC (> 0.6), i.e. EC 1, EC 33, EC 47, EC 89, EC 90, EC 106, EC 183, EC 193, EC 196 and EC 212 were identified. These SSR markers would provide the much needed impetus to cardamom research, globally.

10.5 Conclusion

Generation of genomic resources is of paramount importance particularly in orphan crops, where not much progress has been made in the area of genomics. While a plethora of genomic resources in the form of molecular markers, genetic and physical maps, mapping populations, genomic libraries, genome sequence information, etc., are available for the major crops (cereals, pulses, oilseeds), very scanty resources are available for many agri-horticultural crops.

Crop improvement in these crops is therefore severely lacking behind. With the advances in NGS technologies, we have been able to generate sequence information for a lot of crops. The Zingiberaceae family consists of many species which find immense utility in traditional medicine. Lately, the genomes of many of these species have been sequenced. This has provided the much required momentum to research in these species. One such species is small cardamom, which has recently been sequenced. Through this sequence information, a large number of SSR markers were developed which can be utilized for various purposes for crop improvement in cardamom. Similar approaches in other orphan crops would allow us to unravel the genetic basis of plant metabolism and develop varieties suited for the future.

Funding The authors work in this area is funded by the grant received under the project Indian Council of Agricultural Research-Consortium Research Platform on Genomics, India (project number 1007341).

References

Ahmad A, Wang J-D, Pan Y-B, Sharif R, Gao SJ (2018) Development and use of simple sequence repeats (SSRs) markers for sugarcane breeding and genetic studies. Agronomy 8(11):260. https://doi.org/10.3390/agronomy8110260

Akinola A, Ahmad S, Maziah M (2014) Total anti-oxidant capacity, flavonoid, phenolic acid and polyphenol content in ten selected species of *Zingiberaceae* rhizomes. Afr J Tradit Complement Altern Med 11:7–13. https://doi.org/10.4314/ajtcam.v11i3.2

Anisha CS, Mathew MK, Sasidharan S, Jose S, Varghese RC et al (2020) Diversity analysis of released varieties of Indian cardamom using ISSR markers reveal narrowing genetic base. Indian J Biotechnol 19(4):0972–5849

Anjali N, Sowmya DS, Nadiya F, Sabu KK (2015) Development of EST-SSR markers to assess genetic diversity in *Elettaria cardamomum* Maton. Int J Appl Sci Biotechnol 3(2):188–192. https://doi.org/10.3126/ijasbt.v3i2.12380

Anjali N, Ganga KM, Nadiya F, Shefeek S, Sabu KK (2016) Intraspecific variations in cardamom (*Elettaria cardamomum* Maton): assessment of genomic diversity by flow cytometry, cytological studies and ISSR analysis. SpringerPlus 5(1):1560. https://doi.org/10.1186/s40064-016-3226-x

Awasthi P, Singh A, Sheikh G, Mahajan V, Gupta AP, Gupta S, Bedi YS, Gandhi SG (2017) Mining and characterization of EST-SSR markers for *Zingiber officinale* Roscoe with transferability to other species of Zingiberaceae. Physiol Mol Biol Plants 23(4):925–931. https://doi.org/10.1007/s12298-017-0472-5

Babu Nirmal K, Jayakumar VN, Minoo D, Venugopal MN, Sudarshan MR, Radhakrishnan VV, Backiyarani S, Narayanaswami M, Peter KV, Parthasarathy VA (2012) Genetic diversity and phylogenetic relationships among small cardamom (*Elettaria cardamomum* Maton.) cultivars and related genera using DNA markers. Int J Innov Hortic 1(1):47–56

Beier S, Thiel T, Münch T, Scholz U, Mascher M (2017) MISA-web: a web server for microsatellite prediction. Bioinformatics 16:2583–2585. https://doi.org/10.1093/bioinformatics/btx198

Bhattarai G, Shi A, Kandel DR et al (2021) Genome-wide simple sequence repeats (SSR) markers discovered from whole-genome sequence comparisons of multiple spinach accessions. Sci Rep 11:9999

Biswas MK, Xu Q, Mayer C et al (2014) Genome wide characterization of short tandem repeat markers in sweet orange (*Citrus sinensis*). PLoS ONE 9(8):e104182. https://doi.org/10.1371/journal.pone.0104182

Burkill IH (1966) A dictionary of the economic products of the Malay Peninsula, vols 1 & 2, 2nd edn. Ministry of Agriculture and Cooperatives, Kuala Lumpur, Malaysia

Chakraborty A, Mahajan S, Jaiswal SK, Sharma VK (2021) Genome sequencing of turmeric provides evolutionary insights into its medicinal properties. Commun Biol 4:1–12. https://doi.org/10.1038/s42003-021-02720-y

Chan EWC, Lim YY, Wong SK (2011) Antioxidant properties of ginger leaves: an overview. Free Radicals Antioxid 1:6–16. https://doi.org/10.5530/ax.2011.1.3

Chaudhury R, Malik SK (2004) Genetic conservation of plantation crops and spices using cryopreservation. Ind J Biotechnol 3:348–358

Chaveerach A, Mokkamul P, Sudmoon R, Tanee T (2008) A new species of *Alpinia* Roxb. (Zingiberaceae) from northeastern Thailand. Taiwania 53:1–5

Choudhary R, Chandel KPS (1995) Studies on germination and cryopreservation of cardamom (*Elettaria cardamomum* Maton.) seeds. Seed Sci Biotechnol 23(1):235–240

Christenhusz MJ, Byng JW (2016) The number of known plants species in the world and its annual increase. Phytotaxa 261:201–217. https://doi.org/10.11646/phytotaxa.261.3.1

Cyriac A, Paul R, Anupama K, Senthil RK, Sheeja TE, Nirmal KB, Parthasarathy VA (2016) Isolation and characterization of genomic microsatellite markers for small cardamom (*Elettaria cardamomum* Maton) for utility in genetic diversity analysis. Physiol Mol Biol Plants 22(2):219–229. https://doi.org/10.1007/s12298-016-0355-1

D'Hont A, Denoeud F, Aury JM, Baurens FC, Carreel F, Garsmeur O, Noel B et al (2012) The banana (*Musa acuminata*) genome and the evolution of monocotyledonous plants. Nature 488:213–217

Dong Q, Zou QC, Mao LH, Tian DQ, Hu W, Cao XR, Ding HQ (2022) The chromosome-scale assembly of the *Curcuma alismatifolia* genome provides insight into anthocyanin and terpenoid biosynthesis. Front Plant Sci 13:899588. https://doi.org/10.3389/fpls.2022.899588

Emanuelli F, Lorenzi S, Grzeskowiak L, Catalano V, Stefanini M, Troggio M et al (2013) Genetic diversity and population structure assessed by SSR and SNP markers in a large germplasm collection of grape. BMC Plant Biol 13:39. https://doi.org/10.1186/1471-2229-13-39

Feng S, He R, Lu J, Jiang M, Shen X, Jiang Y, Wang Z, Wang H (2016) Development of SSR markers and assessment of genetic diversity in medicinal *Chrysanthemum morifolium* cultivars. Front Genet 7:113. https://doi.org/10.3389/fgene.2016.00113

Gaikwad AB, Kumari R, Yadav S, Rangan P, Wankhede DP, Bhat KV (2023a) Small cardamom genome: development and utilization of microsatellite markers from a draft genome sequence of *Elettaria cardamomum* Maton. Front Plant Sci 14:1161499. https://doi.org/10.3389/fpls.2023.1161499

Gaikwad AB, Kaila T, Maurya A, Kumari R, Rangan P, Wankhede DP, Bhat KV (2023b) The chloroplast genome of black pepper (*Piper nigrum* L.) and its comparative analysis with related *Piper* species. Front Plant Sci 13:1095781. https://doi.org/10.3389/fpls.2022.1095781

Govindaraj M, Vetriventhan M, Srinivasan M (2015) Importance of genetic diversity assessment in crop plants and its recent advances: an overview of its analytical perspectives. Genet Res Int 431487. https://doi.org/10.1155/2015/431487

Govindarajan VS, Narasimhan S, Raghuveer KG, Lewis YS (1982) Cardamom—production, technology, chemistry and quality. Crit Rev Food Sci Nutr 16:229–326

Gupta PK, Rustgi S, Sharma S, Singh R, Kumar N, Balyan HS (2003) Transferable EST-SSR markers for the study of polymorphism and genetic diversity in bread wheat. Mol Genet Genom 270:315–323

Hajjar R, Jarvis DI, Gemmill-Herren B (2008) The utility of crop genetic diversity in maintaining ecosystem services, pp 261–270

Heryanto R, Syukur C (2021) Study of genetic diversity and relationship of 100 cardamom (*Elettaria cardamomum*) lines based on morphological characters. IOP Conf Ser Earth Environ Sci 752(1):012046. IOP Publishing

Hrideek TK, Radhakrishnan VV, Kuruvilla KM, Mohanan KV (2015) A study on variability of elite landraces of small cardamom (*Elettaria cardamomum* Maton). Indian J Plant Genet Resour 28(3):311–316

Jose S, Mary Mathew K, Rao YS, Kuruvilla KM, Sudarshan MR (2013) Identification of variety-specific ISSR markers in small cardamom (*Elettaria cardamomum* Maton.). J Plant Crops 41(2):233–237

Jose S, Mathew M, Anisha CS, Sasidharan S, Rao YS (2022) Development of an ISSR based SCAR marker to identify small cardamom Malabar (prostrate panicle) variety (*Elettaria cardamomum* (L.) Maton). Sci Hortic 294:110777

Kaila T, Chaduvla PK, Saxena S, Bahadur K, Gahukar SJ, Chaudhury A et al (2016) Chloroplast genome sequence of pigeonpea (*Cajanus cajan* (L.) millspaugh) and *Cajanus scarabaeoides* (L.) thouars: genome organization and comparison with other legumes. Front Plant Sci 7. https://doi.org/10.3389/fpls.2016.01847

Kalia R, Manoj R, Sanjay K, Rohtas S, Ashok D (2011) Microsatellite markers: an overview of the recent progress in plants. Euphytica 177:309–334

Kanehisa M, Goto S (2000) KEGG: Kyoto Encyclopedia of Genes and Genomes. Nucleic Acids Res 28:27–30

Kumari R, Wankhede DP, Bajpai A, Maurya A, Prasad K, Gautam D et al (2019) Genome wide identification and characterization of microsatellite markers in black pepper (*Piper nigrum*): a valuable resource for boosting genomics applications. PLoS ONE 14(12):e0226002

Lapitan VC, Brar DS, Abe T, Redoña ED (2007) Assessment of genetic diversity of Philippine rice cultivars carrying good quality traits using SSR markers. Breed Sci 57(4):263–270

Li HL, Wu L, Dong Z et al (2021) Haplotype-resolved genome of diploid ginger (*Zingiber officinale*) and its unique gingerol biosynthetic pathway. Hortic Res 8:189

Lu JJ, Kang JY, Feng SG, Zhao HY, Liu JJ, Wang HZ (2013) Transferability of SSR markers derived from *Dendrobium nobile* expressed sequence tags (ESTs) and their utilization in *Dendrobium* phylogeny analysis. Sci Horticult 158:8–15. https://doi.org/10.1016/j.scienta.2013.04.011

Madhusoodanan KJ, Pradipkumar K, Ravindran PN (2002) Botany, crop improvement and biotechnology of cardamom. In: Ravindran PN, Madhusoodanan KJ (eds) Cardamom—the genus Elettaria. Taylor & Francis Inc., London, pp 11–68

Malysheva L, Sjakste T, Matzk F, Röder M, Ganal M (2003) Molecular cytogenetic analysis of wheat–barley hybrids using genomic in situ hybridization and barley microsatellite markers. Genome 46:314–322. https://doi.org/10.1139/g02-117

Marks RA, Hotaling S, Frandsen PB, VanBuren R (2021) Representation and participation across 20 years of plant genome sequencing. Nat Plants 7(12):1571–1578. https://doi.org/10.1038/s41477-021-01031-8

Nadukeri S, Umesha K, Naik MK, Srinivasa V, Ankegowda SJ, Hanumantharaya L, Shankarappa KS, Anilkumar GS (2020) Variability studies in small cardamom (*Elettaria cardamomum*) genotypes for growth characters. J Pharmacogn Phytochem 9(6):1346–1349

Nitasha Bhat GM, Nayak N, Vinodraj K, Chandralekha N, Mathai P, Cherian J (2015) Comparison of the efficacy of cardamom (*Elettaria cardamomum*) with pioglitazone on dexamethasone-induced hepatic steatosis, dyslipidemia, and hyperglycemia in albino rats. J Adv Pharm Technol Res 6(3):136. https://doi.org/10.4103/2231-4040.157981

Padmini K, Venugopal MN, Sasikumar B (2000) Performance of hybrids, open pollinated progenies and inbreds of cardamons under nursery conditions. Indian J Agric Sci 70(8):550–551

Pei D, Song S, Kang J, Zhang C, Wang J, Dong T, Ge M, Pervaiz T, Zhang P, Fang J (2023) Characterization of simple sequence repeat (SSR) markers mined in whole grape genomes. Genes 14(3):663. https://doi.org/10.3390/genes14030663

Phadnis S, Peter A (2017) Morphological studies and genetic diversity analysis of cardamom (*Elettaria cardamomum* Maton.) genotypes and *Hedychium coronarium* using RAPD markers. Trends Biosci 10(27):5737–5745

Powell W, Machray GC, Provan J (1996) Polymorphism revealed by simple sequence repeats. Trends Plant Sci 1:215–222. https://doi.org/10.1016/1360-1385(96)86898-1

Prasannan A, Jose S (2021) Assessment of ISSR markers for tagging genetic variability for yield components in small cardamom (*Elettaria cardamomum* Maton). Indian J Agric Res 55(2):212–216

Prasath D, Venugopal MN (2004) Genetic diversity and conservation of cardamom (*Elettaria cardamomum* Maton) in India. Plant Genet Resour Newslett 138:55–60

Preethy T, Murugan M, Mathews N, Kuriakose A (2023) Insights into the genetic diversity of Indian cardamom [*Elettaria cardamomum* (L.) Maton]: for a future research perspective. Plant Genet Resour 21(1):12–18. https://doi.org/10.1017/S1479262123000369

Purseglove JW, Brown EG, Green CL, Robbins SRJ (1981) Spices, vol 2. Longman, New York, USA

Ramanatha Rao V, Hodgkin T (2002) Genetic diversity and conservation and utilization of plant genetic resources. Plant Cell Tissue Organ Cult 68:1–19. https://doi.org/10.1023/A:1013359015812

Ravindran PN, Madhusoodanan KJ (2003) Cardamom: the genus *Elettaria*. Taylor and Francis, London

Ravindran PN, Nirmal Babu K, Saji KV, Geetha SP, Praveen K, Yamuna G (2004) Conservation of spices genetic resources in in vitro gene banks. ICAR project report. Indian Institute of Spices Research, Calicut, Kerala, India, 81 pp

Ren X, Jiang H, Yan Z, Chen Y, Zhou X, Huang L et al (2014) Genetic diversity and population structure of the major peanut (*Arachis hypogaea* L.) cultivars grown in China by SSR markers. PLoS ONE 9:e88091. https://doi.org/10.1371/journal.pone.0088091

Rogers LW, Koehler AM, Crouch JA, Cubeta MA, LeBlanc NR (2022) Comparative genomic analysis reveals contraction of gene families with putative roles in pathogenesis in the fungal boxwood pathogens *Calonectria henricotiae* and *C. pseudonaviculata*. BMC Ecol Evol 22(1):79. https://doi.org/10.1186/s12862-022-02035-4

Saha K, Sinha R, Sinha S (2019) Distribution, cytology, genetic diversity and molecular phylogeny of selected species of Zingiberaceae—a review. Feddes Repert 131:58–68

Sakthipriya M, Sabu KK (2018) Development and cross-genera transferability of ginger EST-SSR markers for cardamom. Curr Bioinform 13(1):95–99

Salgotra RK, Chauhan BS (2023) Genetic diversity, conservation, and utilization of plant genetic resources. Genes 14(1):174. https://doi.org/10.3390/genes14010174

Samir Q, Sasikumar D, Moez AF (2015) Chemopreventive effect of cardamom (*Elettaria cardamomum* L.) against benzo (a) pyrene-induced forestomach papillomagenesis in Swiss Albino Mice. J Environ Pathol Toxicol Oncol 34(2):95–104

Sari D, Sari H, Ikten C et al (2023) Genome-wide discovery of di-nucleotide SSR markers based on whole genome re-sequencing data of *Cicer arietinum* L. and *Cicer reticulatum* Ladiz. Sci Rep 13:10351

Saxena S, Singh A, Archak S, Behera TK, John JK, Meshram SU, Gaikwad AB (2015) Development of novel simple sequence repeat markers in bitter gourd (*Momordica charantia* L.) through enriched genomic libraries and their utilization in analysis of genetic diversity and cross-species transferability. Appl Biochem Biotechnol 175(1):93–118. https://doi.org/10.1007/s12010-014-1249-8

Setyawan AD, Wiryanto W, Suranto S, Bermawie N, Sudarmono S (2014) Comparisons of isozyme diversity in local Java cardamom (*Amomum compactum*) and true cardamom (*Elettaria cardamomum*). Nusantara Biosci 6(1)

Shirasawa K, Ishii K, Kim C, Ban T, Suzuki M, Ito T et al (2013) Development of *Capsicum* EST-SSR markers for species identification and in silico mapping onto the tomato genome sequence. Mol Breed 31:101–110. https://doi.org/10.1007/s11032-012-9774-z

Song X, Yang Q, Bai Y et al (2021) Comprehensive analysis of SSRs and database construction using all complete gene-coding sequences in major horticultural and representative plants. Hortic Res 8:122

Sudharshan MR, Kuruvilla KM, Madhusoodhanan KJ (1991) A key to the identification of types in cardamom. J Plant Crop 18(suppl):52–55

Swarup S, Cargill EJ, Crosby K, Flagel L, Kniskern J, Glenn KC (2021) Genetic diversity is indispensable for plant breeding to improve crops. Crop Sci 61:839–852

Taheri S, Abdullah TL, Rafii MY, Harikrishna JA, Werbrouck SPO, Teo CH, Sahebi M, Azizi P (2019) De novo assembly of transcriptomes, mining, and development of novel EST-SSR markers in Curcuma alismatifolia (*Zingiberaceae family*) through Illumina sequencing. Sci Rep 9(1):3047

Tautz D (1989) Hypervariability of simple sequences as a general source for polymorphic DNA markers. Nucleic Acids Res 17:6463–6471. https://doi.org/10.1093/nar/17.16.6463

Thomas VP, Sabu M (2012) Two new species of *Amomum* (Zingiberaceae) from Western Ghats India. Edinb J Bot 69:313–321

Tyagi RK, Goswami R, Sanayaima R et al (2009) Micropropagation and slow growth conservation of cardamom (*Elettaria cardamomum Maton*). In: Vitro Cell Dev Biol-Plant 45:721–729

Uhlen M, Quake SR (2023) Sequential sequencing by synthesis and the next-generation sequencing revolution. Trends Biotechnol 41(12):1565–1572. https://doi.org/10.1016/j.tibtech.2023.06.007

Varshney RK, Graner A, Sorrells ME (2005) Genic microsatellite markers in plants: features and applications. Trends Biotechnol 23(1):48–55

Wardini TH, Thomas A (1999) *Elettaria cardamomum* (L.) Maton. In: Guzman CC, Siemonsma JE (eds) Plant resources of South-East Asia. No. 13. Spices. Backhuys, Leiden, Netherlands, pp 116–120

Yin Y, Xie X, Zhou L, Yin X, Guo S, Zhou X, Li Q, Shi X, Peng C, Gao J (2022) A chromosome-scale genome assembly of turmeric provides insights into curcumin biosynthesis and tuber formation mechanism. Front Plant Sci 13:1003835

Zhao Z, Zhang H, Wang P, Yang Y, Sun H, Li J et al (2023) Development of SSR molecular markers and genetic diversity analysis of *Clematis acerifolia* from Taihang Mountains. PLoS ONE 18(5):e0285754

Zhou Y, Cao HH, Wang Y, Huang XC, Gu SG, Fu LJ (2020) Application prospect of molecular markers in DUS testing. Anhui Agric Sci Bull 26:15–18

SPRINGER NATURE

GPSR Compliance

The European Union's (EU) General Product Safety Regulation (GPSR) is a set of rules that requires consumer products to be safe and our obligations to ensure this.

If you have any concerns about our products, you can contact us on ProductSafety@springernature.com

In case Publisher is established outside the EU, the EU authorized representative is:

Springer Nature Customer Service Center GmbH
Europaplatz 3
69115 Heidelberg, Germany

The manufacturer's authorised representative in the EU is Springer
Nature Customer Service Centre GmbH, Europaplatz 3, 69115 Heidelberg,
Germany. If you have any concerns regarding our products, please
contact ProductSafety@springernature.com

Printed and bound by CPI Group (UK) Ltd, Croydon, CR0 4YY
01/10/2026
02225916-0020